Time Top 10 Nonfiction Books 2010 *Wall Street Journal*

"Sixteen Books the Book World... ext
B... ...d
B... ...io

"Bro... . It's
not j... able
abou... rch-
ing f... ."

"Cle... s a
thou... o-
log...

"B... r-
tifi... e.
It's... at
see...

"F... t-
ter... he
lib... nt
of someone illuminated in 1849."

—The Millions

"Warm and illuminating."—*Washington Post*

"With grace and authority, Ms. Brox traces the ascendance of artificial light and considers its effect on human culture and psychology."   —*Wall Street Journal*

"Ruminative and curious, Brox excels at discussing the cultural and psychological changes wrought by more and better light."
—*New York Times Book Review*

"[A] brilliant look at not only the creation of artificial light . . . Brox's prose engulfs the reader in the sensations of the eras she describes . . . An astounding read that engulfs the reader in darkness and then metes out light—and illumination—with each chapter. You'll never flip a switch nonchalantly again."
—*Louisville Courier-Journal*

"*Brilliant* is, well, brilliant."—*Portland Press Herald*

"Brox vividly evokes the dark old days when burning lamps left everything in the house coated with soot . . . Brox is a good explainer and here she's at her best."
—*Dallas Morning News*

"Expertly traces the tortuous route to artificial light."
—*New Scientist*

"An enlightened look into the evolution of artificial light . . . written in an easy reading style with lively language and interesting anecdotes that entertain as well as inform . . . Brox covers it all."
—*Sacramento Book Review*

"Superb . . . invariably fascinating . . . With Brox's beautiful prose, this book amply lives up to its title."
—*Publishers Weekly* (starred review)

"Dovetails beautifully with the social history of technology . . . This well-written, well-researched, and thought-provoking book has much to offer."   —*Library Journal*

"*Brilliant* is more than an eloquent and gorgeous history of artificial light; it is a survey of profound experiences long lost to the human senses, imagination and heart . . . All students of literature, history, and art should read *Brilliant*; anyone interested in what it means to be human should read it, too."

—BookBrowse

"[An] odd, enchanting history."—*Orion*

"Invaluable and thought-provoking."—*Booklist*

"Just one of the many pleasures of Jane Brox's sweeping history of human light is its evocation of the wonder and fascination the lowly light bulb roused when it was new, before it became, by virtue of the reverse alchemy of mass production, abundant and déclassé."

—Sylvia Nasar, author of *A Beautiful Mind*

"I'll gladly read anything by Jane Brox on any subject, but her poetic and original retelling of the story of manmade light provides a suitably grand occasion for her superb powers of observation and her intimate, precise, startlingly evocative prose to shine."

—Carlo Rotella, author of *Cut Time*

"Every page contains at least one small marvel, but the greatest wonder is the realization that what [Brox] has illuminated is nothing less than a story of ourselves . . . *Brilliant*, indeed."

—Leah Hager Cohen, author of *Train Go Sorry* and *House Lights*

"*Brilliant* is fascinating in its subject matter, charming in its storytelling and accessible style, and meticulously researched."

—Alan Lightman, author of *Einstein's Dreams*

Brilliant

The
EVOLUTION
of
ARTIFICIAL
LIGHT

Jane Brox

MARINER BOOKS
HOUGHTON MIFFLIN HARCOURT
BOSTON   NEW YORK

First Mariner Books edition 2011

Copyright © 2010 by Jane Brox

www.hmhbooks.com

Library of Congress Cataloging-in-Publication Data
Brox, Jane, date
Brilliant : the evolution of artificial light / Jane Brox.
p. cm.
ISBN 978-0-547-05527-5
ISBN 978-0-547-52034-6 (pbk.)
1. Lighting — History. I. Title.
TH7900.B68 2010
621.32093 — dc22   2009035441

Book design by Brian Moore

Printed in the United States of America

DOM 10 9 8 7 6 5 4 3 2 1

For
DEANNE URMY

and for
JOHN BISBEE

CONTENTS

PROLOGUE

The Earth at Night as Seen from Space

Five hundred years ago, if you could have seen the earth from above, cities, towns, and villages would have appeared nearly as dark as the oak forests. Perhaps glints of light would have leaked through doorways and shuttered windows early in the evening, or a few lanterns would have bobbed down the lanes, but no streetlights would have shone. Within, candles and lamps no brighter than those of Roman times would have lit only a bowl of porridge, a book, a shirtsleeve in need of mending, another. If someone reached for a thread or let out a long sigh, the flame would quiver, and the shadows would quiver, too. Then everything would right itself again. Such small light was precious and meted out sparingly. For much of the evening, people lay in their houses after dousing their cooking fires, sleeping and dreaming away the hours. If by chance on a clear, moonless night they stepped out of their intimate dark and looked up to the heavens, the stars would have been so many that "one could not have put a finger in between them."

Now that our nights are flooded — both outside and in — with insistent light, evidence of our illumination reaches farther into space than any other human thing. On a map of the earth at night as seen from space — made up of a composite of images from satellite photographs taken on nights of the new moon — light blooms across the continents like yeast in warm, sweet water. The edges of land are defined by blazes flaring from the center of cities and waning, though never disappearing, in the suburbs between. Across the United States and western Europe, light wends inland along highways and rivers and stipples the interior foothills, plateaus, and grasslands, diminishing only in the mountains and deserts. Even the interiors of Asia, South America, and Africa — where many people live beyond the reach of electric grids — are salted with small brilliances. The most glaring spots on the map correspond to flagrancy and prosperity rather than density of human habitation: at the moment, the eastern seaboard of the United States is brighter than anyplace in China or India. Only parts of the oceans and the poles appear completely dark.

The story of this increase — just a few centuries old — is one of technology and power, of politics, grievances, and class: the wealthy and powerful have always been the first to acquire new kinds of light and have always had more of it than others. But the story of light is also one of constancy and mystery, of beauty, brilliance, and shadows, and it includes those who continue to use the same types of light now as in centuries past. Even in modern societies, old forms of light live on and acquire new meaning: an open flame — brought to life by a gesture of the hand, extinguished with a breath — has always been much more than simply a utilitarian tool, for it holds the power to fix our gaze and free our thoughts, and it lies at the heart of our making, thinking, and dreaming. "We are almost certain that fire is precisely the first object, the *first phenomenon*, on which the human mind *reflected*," wrote Gaston Bachelard.

The story is in the light — in its utility and beauty — and

also in the way its increase has altered life by granting more working hours in the day and creating a night that is no longer impenetrable, no longer a void, a night easily traveled through and expansive with free time. What does it mean to have new hours for the human spirit? How have wealth and privilege shaped those hours? And what are the consequences for those who continue to live without modern light? Have our bodies and minds adapted to a world in which the tides of the day are lost and the stars appear to have vanished? What of pumas, loggerhead turtles, and cockleburs? How has the way we think about light changed now that we've left the solitary lamp behind, now that we've bound ourselves to the grid? And can an understanding of the way those in the past adapted to their own new sources of light help us to better illuminate our future?

PART I

Of time that passes by burning . . .

—GASTON BACHELARD,
The Flame of a Candle

1

LASCAUX: THE FIRST LAMP

ALTHOUGH FIRE HAS BLAZED in hearths and flared from pine torches for half a million years, the earliest known stone lamps — fashioned by Ice Age humans during the Pleistocene — are no more than forty thousand years old. Their quiet flames shone more weakly than those of our candles, but they were cleaner than torchwood and easier to guard and tend. Often the lamps were merely unworked flat slabs of limestone, or limestone with natural cavities for the nubs of tallow — animal fat — that had to be replenished every hour. Some were roughly carved and their reservoirs carefully shaped with sloping sides so that the melted fat could be poured off without drowning the lichen, moss, or juniper wicks. Since limestone is a poor conductor of heat, there'd have been no need to carve a handle: people could hold the lamp in the palm of their hands. Except that the cups are charred, they could be mistaken for small mortars or grinding stones.

Archaeologists have discovered such stone lamps overturned near open hearths and among cooking tools and spearpoints in shallow rock shelters. They've also unearthed them far from settlements, deep in the caves of what is currently southern France, caves that are now famous — La Mouthe, Lascaux — for there isn't anything more beautiful than what Ice Age humans made by such light. Eighteen thousand years ago, while above them herds funneled through valleys on their way to the plains near the coast, people ventured far beyond the reach of day — working their way down stone corridors and twisting through narrows — to draw from memory on the limestone walls and ceilings. Sometimes their works extend higher than human reach: a man would have had to stand on scaffolding or upon a rock protruding from a wall to make marks with his hands and with bristles dipped in manganese and iron oxide. More often, the artist held the pigment in his mouth and blew it onto the cave wall to make a mark. He also blew through hollowed-out bones. Concentrated marks one after another produced the sturdy outline of an animal, while a more diffuse spray colored a flank or back. In places details are certain and fine. Elsewhere the marks are suggestive: four streaks make a cat's head. At times the contours of the wall stand for the back of a horse, a small protuberance for an eye. The artists understood how to place a leg or draw the turn of a head to create a sense of visual depth in their work.

In the chambers of Lascaux, black and brilliant animals swirl, eddy, and flow toward the deepest reaches of the cave: Galloping horses and horses superimposed on horses, a great red and black horse, a horse with a turned-back foot, a horse rolling on the ground, traces of a painted equid. A black stag, swimming stags, a fallen stag, a stag with thirteen arrows. A great stag and horse with merged outlines. A headless equid drawn in red. Two bison, the head of a bison, the head and horns of a cow, a red cow painted on the ceiling. The solitary head of a bull in the Hall of the Bulls. Panel of the Musk Ox,

Panel of the Ibexes, Niche of the Felines. Wounded, grazing, fleeing, young: "The iconography of this cave," said archaeologist Norbert Aujoulat, "is, above all, a fantastic ode to life." Everything was contingent on the herds: food and clothing (needles and awls were carved from bones, while tendons provided thread and binding), as well the tallow in the lamps.

There's no evidence that Ice Age humans used more than a handful of lamps as they drew, and if carbon dioxide had built up in the chamber — as it often does in the still air of deep limestone caves — they might have had trouble keeping even their few lamps lit. It's likely they saw only a small portion of their work at any one time, that it receded in darkness behind them and lay in shadow above them: "Achieving full and accurate color perception of the cave images along a five-meter-long panel," notes French archaeologist Sophie de Beaune, "would require 150 lamps, each of them placed 50 centimeters from the cave wall." So the artists couldn't have perceived the reds, yellows, and blacks of their own marks as clearly as we moderns can under the incessant glare of electric bulbs or in contemporary color photographs of the friezes and panels.

To reach the farthest chamber of Lascaux, it's likely a man had to snuff out his light, lower himself down a shaft with a rope made of twisted fibers, and then rekindle his lamp in the dark so as to draw the woolly rhinoceros, the half horse, and the raging bison there. A long spear transfixes that bison, and entrails pour from its side. Beneath its front hooves lies the one painted man in all of Lascaux: prone, spindly, wounded, disguised behind a bird mask. And below him, until its discovery in 1960, lay a spoon-shaped lamp carved of red sandstone. It differs from the others in more than the nature of the stone and its shape. (The handle was essential because sandstone conducts heat efficiently, and it would have been impossible to hold the lamp without it.) The lamp possesses a refined beauty: its maker created a perfectly symmetrical bowl, polished the sandstone smooth, and incised the handle with chevrons. Perhaps it was used for ceremonies, though that can't entirely be known. Hold it again as it once was held, and the animals will emerge out of darkness as you pass. Nothing stays still. Shadows nestle in the cavities; a flicker of light across pale protruding rock turns a hoof or raises a head. One shape recedes as another emerges, and everything lingers in the imagination.

Light as it would be for ages to come: light, its limits, and then the dark. Over time, lamps were fashioned out of shells, then pottery shaped like shells or slippers, and there were gradual improvements in the design: some bear turned-over lips on their terra cotta cups, which prevented spills. The cloth or rope wicks lay horizontally within wick channels shaped like thick spouts — perhaps suggested by the flutes of shells — which helped the oil to climb the wick and keep the flame steady. Ancient Greek and Roman lamps had enclosed reservoirs, which protected the oil from dirt or flies and guaranteed a little safety, but the flame itself was unguarded by glass.

It is believed that the Romans might have fashioned the first beeswax candles, which gave a fragrant, clear, steady flame and burned so evenly they were eventually used to divide time into hours. The ninth-century Anglo-Saxon king Alfred the Great wished to "render to God, with a good heart, the fourth part of the service of his body and of his mind, both by day and by night." So as to tell accurate time in the dark or in the rain, he ordered that beeswax equal to seventy-two pence in weight be made into six candles, each twelve inches long. He needed to prevent drafts from affecting the burning time of the candles, for "the violence of the winds blew too much upon them

. . . day and night without ceasing through the doors of the churches and the windows, and the chinks and holes in the woodwork, and the many rifts in the walls, and the thin tents." To do so, he "ordered a lantern to be well made of wood and ox-horn, for the horns of oxen, when white and planed down to a thin sheet, are as clear as glass. . . . And when this device had been so executed, six candles, one after another, burned for twenty-four hours without intermission, neither too quickly or too slowly. And when they went out others were lighted."

Rare and costly beeswax was long the province only of the Roman Catholic Church and the wealthy. Most other people depended on fat they pressed or rendered from animals, fish, or vegetation near at hand: manatees, alligators, whales, sheep, oxen, bison, deer, bears, coconuts, cottonseed, rapeseed, and olives, the chosen oil of the Mediterranean. In England tallow candles from domestic herds provided the main source of light. The highest-quality candles contained a large portion of hard, white mutton tallow, while softer beef tallow made a taper of lesser quality. Poor people couldn't be fussy about their tallow and would use almost any household grease available for their lights, which were most often made of rushes that had been gathered from the marshes in late summer or fall. The work of making such lights was usually reserved for children and the old, who soaked the rushes and peeled away the outer skin. They dried the inner pith in the sun, then repeatedly dipped the rush in melted fat. Rushlights were frail and slim — "an object like the ghost of a walking-cane," wrote Charles Dickens, "which instantly broke its back if it were touched." A simple iron pincer held the rush at a slant, for upright it consumed itself too quickly. A well-made two-foot rushlight would burn shy of an hour.

Light, it seems, could be gained from any viable thing at

hand. In the West Indies, the Caribbean, Japan, and the South Sea Islands, people saw by the light of numerous fireflies, which they captured and kept in small cages. South Sea Islanders skewered oily candlenuts on bamboo to make torches, while those on Vancouver Island placed a dried salmon in the fork of a stick and lit it. Shetland Islanders caught, killed, and stored storm petrels by the thousands. The petrel, it's said, was named after Saint Peter, because it seems to walk on water as it feeds: a sea bird, full of buoyant, insulating oil. When the islanders needed a lamp, they'd affix a petrel carcass to a base of clay, thread a wick down its throat, and set it alight.

The first American colonists — possessing no domestic herds in the early years of settlement, but being surrounded by abundant woodlands — often used pine knots, called candlewood, for their lights. The knots smoked heavily and dripped pitch, so they were usually placed in the corner of a fireplace or on a stone slab. Wood splinters set in iron pincers provided portable lights. Even after herds were established in the colonies, poorer people continued to use candlewood, as did rural families: "It was said that a prudent New England farmer would as soon start the winter without hay in his barn as without candle-wood in his woodshed."

New Englanders sometimes made fragrant candles from the waxy outer coating of bayberries, which they rendered by boiling the berries. They also made use of deer, moose, and bear fat, although once they established herds of sheep and cattle, they used the fat of their domestic animals as well. Women spent long hours painstakingly dipping candles — "a serious undertaking . . . sevenfold worse in its way even than washing-day," claimed Harriet Beecher Stowe. "A great kettle was slung over the kitchen fire, in which cakes of tallow were speedily liquefying; a frame was placed quite across the kitchen to sustain candle-rods, with a train of board underneath to catch the drippings." The day could not be too warm, or the quality of the candles would suffer. The tallow had to be "cut very small, that it may be speedily dissolved; for otherwise it would be liable to burn or become black, if left too long over the fire." The wicks couldn't be dipped too quickly, or the candles would be brittle. After the first three dips, "water, proportionate to that of tallow, [was] poured in for precipitating the impure particles to the bottom of the vessel." It could not be done sooner, "as the water, by penetrating the wicks, would make the candles crackle in burning, and thereby render them useless." Afterward, the candles had to be cooled slowly, or they would be likely to crack. They softened in warm weather and, being made of animal fat, spoiled on the shelf over time. They had to be stored where the mice and rats couldn't get at them.

In later years, women used tin or pewter molds to make candles. Their work then was simpler and quicker, though still laborious, for a farm wife would have to make hundreds of candles to last for a winter of meager light. Historian Marshall Davidson notes that "even the best-read people remained sparing with candlelight. In his diary for 1743 the Reverend Edward Holyoke, then president of Harvard, noted that on May 22 and 23 his household made 78 pounds of candles. Less than six months later the diary records in its line-a-day style, 'Candles all gone.'"

Unlike the paraffin candles of modern times, tallow candles were not easy to keep lit. Not only did they soften in warm weather, but they also burned unevenly and lost their brilliance as they burned. To maintain more than a few at any one time required constant work: each would have to be snuffed — that is, the charred wick had to be trimmed — and rekindled at least every half-hour to be kept from guttering. (Guttering occurs when the melted wax channels down the side of the candle, which makes the taper burn unevenly and causes the flame to flicker.) A draft would misshape and often douse a candle. If it wasn't properly extinguished, it would give off excessive smoke and an acrid stench, which was all the more problematic in well-to-do households, where many candles might be extinguished at once. Jonathan Swift gave extensive advice to servants concerning the dousing of candles:

> There are several Ways of putting out Candles, and you ought to be instructed in them all: You may run the Candle End against the Wainscot, which puts the Snuff out immediately: You may lay it on the Floor, and tread the Snuff out with your Foot; You may hold it upside down until it is choaked with its own Grease; or cram it into the Socket of the Candlestick: You may whirl it round in your Hand till it goes out: When you go to Bed, after you have made Water, you may dip the Candle End into the Chamber-Pot: You may spit on your Finger and Thumb, and pinch the Snuff until it goes out: The Cook may run the Candle's Nose into the Meal Tub, or the Groom into a Vessel of Oats, or a Lock of Hay, or a Heap of Litter. . . . But the quickest and best of all Methods, is to blow it out with your Breath, which leaves the Candle clear and readier to be lighted.

As for lamps, even with tallow of the highest quality, they needed frequent cleaning to work well. Tallow, being thick, had trouble climbing up the wick — often nothing more than a twisted rag in poorer households — which had to be pulled up from time to time and trimmed. If the fire was starved of fuel, it would produce a thin, smoky flame, though given too much, it would smoke as well. And it smelled gamy: "stinking tallow," Shakespeare called it.

In every century, those who had easy access to an ample fuel supply could enjoy adequate light, as did the wealthy everywhere, who also brightened their homes and halls by making use of precious mirrors to magnify the flames and who could be profligate with their beeswax. "At the Court of Louis XIV of France no candle was ever re-lighted and the ladies-in-waiting made quite a good thing out of selling, as their perquisite, the candle ends of expensive wax," notes historian William O'Dea. "This seems to have been the custom in other royal households." But for those who had to buy candles, the cost was dear: "In the middle of the fifteenth century in Tours, a laborer had to work half a day to earn enough for a pound of tallow. And wax was priceless."

The lamp was one thing; lighting it was another, especially before the invention of the safety match in the nineteenth century. The earliest intentional fires were started with sparks made by striking flint against iron pyrites, or from the friction between hardwood and softwood, for which the fire builder might lay a hardwood stick set with drilled holes on the ground or across his knees, then insert a softwood stick into one of the holes and twirl it steadily — perhaps for less than a minute on a day with no rain — until the abrasion created enough heat to start the wood smoldering. Once he saw smoke rise, he would throw crushed dry leaves on it, cup the smoldering leaves with his hands, and blow the smolder into a blaze. Then he'd turn the fire over onto a small pile of twigs and leaves. It would be easiest to start with good dry sticks, which were often much cherished. Of the Karankawa Indians of Texas it is said: "Their fire sticks they always carried with them and kept them carefully wrapped in several layers of skins tied up with thongs and made into a neat package; they were thus kept very dry, and as soon as the occasion for their use was over they were immediately wrapped up again and laid away."

In eighteenth-century Europe, getting a flame was hardly

any easier. The tinderbox found in almost all kitchens would have contained fire steel, flint, and tinder — usually charred linen. To make a fire by striking flint against steel and setting off sparks, which were aimed toward the charred cloth, fed with more tinder, and fanned to a flame, was an ordinary task that could be accomplished quickly on a dry day in broad light, though on "a cold dark frosty morning when the hands are chapped, frozen and insensible," wrote one sufferer, "you may chance to strike the flint against the knuckles for some considerable time without discovering your mistake."

Once gotten, fire was carefully guarded, and many households maintained some glowing embers in the hearth. If the fire went cold, a child would be sent to a neighbor's with a pail or shovel to fill with live coals. James Boswell, author of *The Life of Samuel Johnson*, wrote of the consequences of losing one's light:

> About two o'clock in the morning I inadvertently snuffed out my candle, and as my fire was long before black and cold, I was in a great dilemma how to proceed. Downstairs did I softly and silently step into the kitchen. But, alas, there was as little fire there as upon the icy mountains of Greenland. With a tinder box is a light struck every morning to kindle the fire, which is put out at night. But this tinder box I could not see, nor knew where to find. I was now filled with gloomy ideas of the terrors of the night. . . . I went up to my room, sat quietly until I heard the watchman calling 'past three o'clock'. I then called to him to knock at the door of the house where I lodged. He did so, and I opened to him and got my candle re-lumed without danger.

Sometimes the lack of a candle could be deadly. Historian Jane Nylander uncovered the record of an "unfortunate man staying at a tavern in New Haven in June 1796 [who] 'was going to bed without a light . . . [and] opened the cellar door instead of a chamber door, and falling down the cellar steps fractured his Scull, of which he expired the next morning.'" But also the danger of fire from an open flame never ceased. In truth, as cities grew larger, entire districts of tightly packed wooden houses were at the mercy of an overturned lamp, a stray cinder, a child careless with a candle. One eighteenth-century writer noted, "The English dwell and sleep, as it were, surrounded with their funeral piles."

Such danger might be reason enough to send the children to bed in the dark, but more likely it was done for economy's sake. Before the advent of mineral oils in the nineteenth century, all fuel could also be used for food. John Smeaton, in his account of building the Eddystone Lighthouse off the coast of Plymouth, England, said that he "found it a matter of complaint through the country — that the light keepers had at various times been reduced to the necessity of eating the candles."

In the worst of times, many saw only by the light of their cooking fires, or by dint of one candle or lamp at the center of a table, which they rarely lit before darkness fell. The poorest people might have no light at all. So a glimmer for a task, for an hour, for supper in winter. Farmers might repair their tools or carve new ax handles by lamplight. Women mended and stitched. It was hardly enough for precise work: "A French *Book of Trades* in the thirteenth century forbade gold and silversmiths to work [after dark], for 'light at night is insufficient for them to ply their trade well and truly,'" notes historian A. Roger Ekirch. But what constituted "dark" wasn't often clear: "From Easter to Saint-Rémi, tannery workers set the rising and the setting of the sun as the limits of the working day for summer, and for winter, the moment when there was not enough light to distinguish a denier [a small coin] of Tours from a denier of Paris."

In a time when labor was often ceaseless during the day, the constrictions of the night could be welcome. According to Cyril of Jerusalem, "A servant would have had no rest from his masters, had not the darkness necessarily brought a respite. And often after wearying ourselves in the day, how are we refreshed in the night." The church, however, deemed night not only as a time of rest but also as a time for prayer and for the soul's reckoning: "And what [is] more helpful to wisdom than the night?" asked Cyril. "And when is our mind most attuned to Psalmody and Prayer? Is it not at night? And when have we often called our own sins to remembrance? Is it not at night?" Beyond rest and prayer, in the dimly lit interiors, in the close and crowded quarters of earlier times, people may have even found a little freedom within the confines of their homes, for the dark affords its own kind of privacy: no one and no thing can be fully seen.

Still, people devised ways to increase what little light they had. Sometimes they would focus and magnify their lights by setting a water bottle in front of a flame. In European villages, women would gather at one cottage in the evening and position themselves around a raised lamp that had been surrounded with globes of tinted blue water. (Women in cold countries used snow water.) The color, it was said, tempered the glare. Though all kinds of close work was done by such light, this was called a lacemaker's lamp. The workers gathered "in orderly rows," Gertrude Whiting explained, "the best lacemakers on the highest stools nearest the lamp or candle-stand. Thus, we are told, some eighteen workers can be accommodated, the outer row of stools or chairs being lower to catch the falling rays of light shed from the pole-board. This graded arrangement is spoken of as *first, second* and *third lights."* Third light would have been particularly ghostly: the women facing the inky backs of their companions, gleaning light from the diffuse rays that fell from above or between those in front of them. It illuminated little more than their hands and work.

Is it any wonder that in good weather women sat at the door of their homes and sewed, mended, or made lace in broad daylight? Although in seventeenth-century Amsterdam, large windows lightened the interiors of homes and showed up the dirt in the corners as never before — spurring housewives to sweep and scrub all that much harder — rooms were still consumed by shadows. In Vermeer's *The Little Street*, the inside of a home glimpsed through glass windows appears dark in day, as it does through the open door where a woman in a white cap sits, intent on the white work in her lap. She's framed in whitewash, then in sturdy, centuries-old brickwork, which has settled and cracked and been patched. The high façade makes the Dutch street seem akin to the shallow rock shelters of the last Ice Age, where women — bent over sinew, stone, and bone — also sat in the open, patiently tending to fleeting life.

TIME OF DARK STREETS

LIGHT — SO PRECIOUS WITHIN — was even rarer on the streets of the cities, towns, and villages of the past, for before the seventeenth century, street lighting was almost nonexistent everywhere in the world. A fourth-century inhabitant of the Syrian city of Antioch claimed, "The light of the sun is succeeded by other lights. . . . The night with us differs from the day only in the appearance of the light." And geographer Yi-Fu Tuan notes that in China, "Hang-chou boasted a vigorous night life along the crowded Imperial way before the Mongols invaded the Sung capital in A.D. 1276." But other Chinese cities were dark except during the New Year and on the emperor's birthday, when torches lined the roads and the skies flared with fireworks. Renaissance Florence had no streetlights, nor did imperial Rome, of which Jérôme Carcopino wrote:

No oil lamps lighted [the streets], no candles were affixed to the walls; no lanterns were hung over the lintel of the doors, save on festive occasions when Rome was resplendent with exceptional illumination to demonstrate her collective joy, as when

Cicero rid her of the Catilinarian plague. In normal times night fell over the city like the shadow of a great danger.... Everyone fled to his home, shut himself in, and barricaded the entrance. The shops fell silent, safety chains were drawn across behind the leaves of the doors; the shutters of the flats were closed and the pots of flowers withdrawn from the windows they had adorned.

In Europe during the Middle Ages, the close of day was un-mistakably announced with the clanging and groaning of bells. Bells were always ringing from ramparts and cathedral tow-ers, from the belfries of convents, monasteries, and country churches — to warn of invasions, fires, and thunderstorms; to announce the celebration of marriage, the arrival of royalty, and the impending death of a parishioner; and after death, to solicit prayers for the departed soul. Their sounding shaped time into holy hours — matins, lauds, prime — and marked the ordinary — the start of work, the opening of markets, the re-spite of noon. Come dusk, the vespers bells rang, calling for the holy office of the lights, when the candles and torches of the churches were lit. Vespers, meaning "evening star," the word itself dying on a silky whisper: hour for prayers of thanks-giving, and for prayers to the Virgin Mary, since people be-lieved that the Annunciation took place in the evening.

Soon after, the curfew bell tolled, often more than a hun-dred times. In the early Middle Ages, it sounded just after dusk; in later centuries, especially in winter, it rang several hours after sunset. But always it held an unwavering meaning: in a time before street lighting or organized police forces, the only way to maintain order was to strictly control people's comings and goings, so at curfew all the day's labor stopped. Blacksmiths lay down their bellows, and goldsmiths ceased beating out metal. Trading halted in the markets, and the cries of butchers and fishwives subsided. The sounds of clinking harnesses, creaking wagons, and the plodding tread of oxen decayed into silence as almost everyone — per order of the authorities — returned to their dwellings, locked their doors, and shuttered their windows.

If inhabitants of fortified cities and towns found themselves beyond the gates at the sound of curfew, they made true haste, since officials, to prevent intruders from entering under the cover of dark, locked the perimeter gates. Anyone caught beyond them risked being fined or shut out for the night. Such a practice persisted in some places even into the eighteenth century: "About half a league from the city [of Geneva]," Jean-Jacques Rousseau attests, "I hear the retreat sounding; I hurry up; I hear the drum being beaten, so I run at full speed: I get there all out of breath, and perspiring; my heart is beating; from far away, I see the soldiers from their lookouts; I run, I scream with a choked voice. It was too late."

Not only were gates closed; in order to prevent vandals from running freely through the streets, officials laid chains across the roads, "as if it were in tyme of warr." The city of Nuremberg, notes A. Roger Ekirch, "maintained more than four hundred sets [of chains]. Unwound each evening from large drums, they were strung at waist height, sometimes in two or three bands, from one side of a street to the other . . . [and] Paris officials in 1405 set all the city's farriers to forging chains to cordon off not just streets but also the Seine." In some cities, residents, once home, were required to give their keys over to the authorities: "At night all houses . . . are to be locked and the keyes deposited with a magistrate," a Paris decree of 1380 charged. "Nobody may then enter or leave a house unless he can give the magistrate a good reason for doing so." Cooking fires, often the only interior light many could afford, were ordered extinguished soon after the evening meal, since among the innumerable night fears in the huddled wooden-and-thatch world of the Middle Ages was that of conflagration. "Curfew" comes from the Old French *covre-feu*, meaning "cover fire."

Yet even with such strict regulations, and in spite of all the tolling bells and clanking chains, the close of day was not always an iron hour. The absolute enforcement of curfew would have been impossible, since the night watch was often all that stood between order and disorder in the dark, and watchmen weren't at their posts voluntarily. In many European cities and large towns, all households were required to contribute a man between the age of eighteen and sixty to the watch, and neither widows nor clergy were excepted from the ordinance: they had to sponsor an eligible man from another household. Unpaid, unarmed (save for a trumpet and banner), and having worked all day as laborers, goldsmiths, or cloth makers, the standing watch kept a lookout for fire or invasion at the towers and gates, having climbed to their posts on ladders, "whose feet in many towns were protected by a locked barrier. Thus, the watchers . . . would not be tempted — or more precisely would not be able — to abandon their post under cover of darkness. Installed in sentry boxes, suffering in winter from cold and bad weather, they waited more or less patiently for night to pass." A rear watch spent the night patrolling the streets, listening for trouble and questioning anyone found abroad. They had the additional duty of checking on the standing watch to make sure one or more of them hadn't dozed off or returned home.

All watchmen had the authority to arrest and imprison those out in the night without just cause, though they might be a little lax in the first few hours after curfew, especially in times and places that were relatively free from strife. The taverns, though ordered closed, might have stayed open so workmen could stop in for a drink or two before returning home. In small towns and villages, people visited other households to talk by the light of the hearth. Bakers worked their ovens so they could have bread ready for the break of day. And the night had its own tradesmen who were about then — ragpickers, manure and night soil collectors — with their furtive scrapings and footsteps. But as night deepened, the streets mostly belonged to vandals, footpads, and other thieves, and anyone abroad in the later hours except those with a legitimate purpose — midwives, priests, or doctors called out to emergencies — was regarded as a "nightwalker" and subject to interrogation.

Since the watch — and any travelers abroad — had no stationary street lighting to help them, what little light shone on the streets was portable. The torches and lanterns carried by the watch not only illuminated their way but also made them visible to others and recognizable as enforcers of order. Since any travelers without lights would have had an advantage — they could see the watch but could not be seen — anyone on the streets after dark was also required to carry a lamp or torch. Leicester, England: "No man [may] walke after IX of the belle be streken in the nyght withoute lyght or without cause resonable in payne of impresonment." The city of Lyon: "Let no one be so bold or daring to go about at night after the great *seral* of Saint Nizar without carrying lights, on pain of being put in prison and of paying sixty sous of Tours each time he is found to have done so."

The wealthy — whom watchmen could distinguish from a distance by their dress — always traveled with servants to hold their lanterns and with guards to defend them. They were also exempted from nighttime restrictions that others were subject to. For instance, in many cities night travelers were forbidden to wear hoods or cloaks, and they could not carry weapons or gather in groups of more than three or four.

Almost everyone gladly left the streets to the thieves, the scurrying of rodents, and the lingering smells of the day — rotting food, old straw, and horses' sweat and dung: "It has been said in describing the conditions of the age of dark streets that everybody signed his will and was prepared for death before he left his home." Women would have been particularly vulnerable in the night, and any women on the streets after nightfall, save for midwives, would have been deemed to be prostitutes.

People who had to travel hoped that their business would coincide with a clear night and a full moon, which thieves often avoided. The full moon also gave travelers enough light to see the outlines of the landscape and the road ahead. The eye functions differently at night than during the day. In the dark, people see with their retinal rods rather than their retinal cones, and complete adaptation to night vision takes a full hour. Even then, human sight is much less acute at night, and the eye can't distinguish color. On a night with no moon or heavy cloud cover — a lantern or torch lighting the way only directly ahead — travelers relied on their other senses. Most knew the country intimately by day, and such familiarity would have helped in the dark. Although they could not see landmarks, they could orient themselves by the feel of the road underfoot — the gravel crunching with each step or the give of soft sand; by the sound of the wind soughing through the trees or rushing across an open field, or by church bells, falling water, or bleating sheep; by the smell of hay or freshly cut wood. Anything light-colored helped — a pale horse, a sandy path, snow. Still, people had to negotiate the curfew chains or the logs that were sometimes placed across the streets as barriers. On the uneven, muddy roads, they fell off bridges and into canals and coal bins; they stumbled over cobbles; they tripped on woodpiles and stones.

In an age of scarce and rarely squandered light, any substantial illumination at night would have been imbued with great meaning. At times it signaled a crisis: during conflagrations or conflicts, city officials required citizens to muster their lamps and candles as an aid to defense or firefighting. At other times, it signaled power: when royalty arrived in a city, they were often ushered in with displays of torches along the streets and on rooftops, or with bonfires: "On the twenty-sixth day of April 1430, the authorities of Paris had great fires lit, just as at Saint John in the summer . . . and informed the people that it was for the young King Henry, who proclaimed himself king of France and England, who had landed at Boulognes, he and a great horde of mercenaries, to fight the Armagnacs, who were nothing to him." The church also marked its holy days with fire and used light extravagantly in its buildings. Of St. Mark's in Rome on Christmas Eve, one onlooker remarked, "A man would thincke it all on fire." While such light reinforced the church's eminent place in society, candlelit processions through the streets and squares also imbued those times with a sense of solemnity and mystery.

Even in the heart of a bustling city, night must have retained its age-old feeling of enormity, with the stars distinct above and people hiding in their homes. Over time, though, as cities grew and commerce between and within them increased, daily life inevitably extended more and more into the dark hours.

By the late 1600s, authorities in large European and several American cities began to require householders to hang a lamp or place a candle on their street-facing windowsills for a few hours after winter sunset and during the dark of the moon. Like the lights required of travelers, the sill lights were meant to help the authorities. The lamp on the sill was also "a lamp that *waits*," notes Gaston Bachelard. "It watches so unremittingly that it *guards*." And it was regarded in return. There would always be something of the cold taste of order in public lighting.

The times and days set for lighting the lanterns remained variable for centuries — changing with the seasons, with the lunar phases, and with days of religious observance. Eventually, officials issued detailed schedules. In 1719, for example, a Paris district commissar required the following: "On 1 December a half candle (⅛ pound) is to be lit. From 2 to 21 December inclusive, whole candles (¼ pound) are to be used. On 22 and 23 December no candles are to be lit. On 24 December, Christmas Eve, twelve-pound candles are to be burned. From 25 to 27 December, no lighting is to be used at all."

Oftentimes citizens resented the obligation. In New York, for instance, "the magistrates — remarking on 'the great Inconvenience that Attends this Citty, being A trading place for want of Lights,' — ordered that every house have a light 'hung out on a Pole' from an upper window 'in the Darke time of the Moon.' When homeowners objected to the expense, the magistrates retreated to a requirement that only every seventh house need present 'A Lanthorn & Candle,' and only in winter, the cost to be shared by the owners of the other six." The duty was undertaken reluctantly, not only because of the expense but also because it was a fussy task — lights had to be constantly tended to keep them from guttering, smoking, or dying out. And if by chance a watchman spied a cold lantern, he would rouse the one responsible for it and make him tend it.

Those first lanterns and candles hardly figured in the dark. At best they were unsteady and faint, barely protected from wind and rain, and easily put out with a stick or stone. But they also marked the beginning of a new conversation with night, offering a little more freedom and time — maybe for work, maybe for the counter life that the night always offered: a chance for pleasure and the risk of transgression. And one after another down the streets, like channel markers, they staked the human community in the dark: here, here, here, here.

Light always seems to beget more light. As the nightscape grew livelier with comings and goings, with the sound of human voices leaking from taverns and coffeehouses, which had become common by 1700, with their late hours and their offerings of stimulants — tea and chocolate as well as coffee — keeping order became a more complex task for the authorities, and they required more, and more dependable, light to help them. On the corners of the largest streets in Boston, the night watchmen kept iron fire baskets filled and burning (there would not be streetlamps there until the late eighteenth century), and in London, Paris, New York, Turin, Copenhagen, and Amsterdam, authorities erected stationary streetlights to replace residents' sill lamps. Maintained by the cities and paid for with taxes, the lights not only shone more frequently in the winter but also were often lit during the summer and during all phases of the moon.

Even so, to the English writer William Sidney, the streetlamps in eighteenth-century London were "totally inadequate to dispel the Cimmerian gloom in which London was shrouded in the winter months." Sidney continued:

The light, such as it was, was derived mainly from several thousands of small tin vessels, which were half filled with whale oil of the worst quality that could possibly be procured, supplied with bits of cotton twist for wick and enclosed in globes of semi-opaque glass . . . and served to shed a faint glimmer of light, or rather to make the darkness visible at street corners and crossings from sundown to midnight, when they were religiously extinguished, if they had not in the meantime rendered this duty unnecessary by extinguishing themselves.

The lamplighters, who by then were employed to tend the lamps, were, according to Sidney, "greasy clodhopping fellows. . . . A distinguishing characteristic of these lamplighters was that of invariably spilling the oil upon the heads of those who passed them while they stood upon their ladders, and occasionally breaking a head by dropping a globe." The French writer Louis-Sébastien Mercier likewise complained about the lamplighters of Paris: "Another thing; they might as well keep some sort of watch on the lamplighters. These fill the lamps, as they call it, nightly; actually they allow so little oil that by nine or ten o'clock half are already out; only an occasional and distant glimmer reminds you of how the streets should look."

Officials in a few cities believed that streetlights actually abetted criminals. Yi-Fu Tuan notes, "Cautious citizens in Birmingham did not want to experiment with new lighting; they believed the crime rate in their city was lower than London's because their city was so dark." Likewise, officials in Cologne believed "that as the fear of darkness vanished, drunkenness and depravity would increase." They argued further that if street lighting became common, festive and ceremonial lighting would lose some of its wonder. But in most major cities, those in charge thought otherwise and attempted to light as many streets as possible, for if light was the mark of authority, dark neighborhoods would be uncontrollable, full of troublemakers who'd been chased away from well-lit streets. For this reason, the widespread practice of lantern smashing was punishable by imprisonment or worse: "In Vienna in 1688, authorities threatened to cut off the right hand of anyone caught damaging a street lantern."

Lights and more lights had unintended consequences, and it isn't always easy to distinguish the ways that they helped from the ways that they hindered. Under the lanterns, the streets grew rowdier. Frequenters of taverns didn't have to sit on the same stool all night; they could now make their merry way more easily from the Crown and Anchor to the White Horse and then to the Black Crow. And the pools of light interspersed with shadows were a great help to prostitutes, who in the Middle Ages had been largely confined to brothels and bathhouses. They now stood under the streetlamps to entice their customers, then quickly withdrew into the shadows for their assignations.

The night, always silent of the hawkers' cries for apples, cabbages, herring, and mutton that sounded through the day, was now filled with the calls of torchbearers — known as linkmen or linkboys, for the "links," or torches, they carried. They roved

the streets after ten at night, crying, "Here's your light." After supper is the best time for this cry, and these fellows go calling and answering one another, all night long, to the great prejudice of those whose bedrooms face on the street; they are to be found in clusters at the door of any house of entertainment. . . . The man lights you to your door, to your bedroom — if seven flights up, no matter — and this aid is of value when perhaps you keep no servant . . . a plight not rare among smart young men, most of whose money goes in coats and theatre tickets. These wandering lights are a protection, besides, against thieves; and are in themselves almost as good as a squad of watchmen. . . . They are, in fact, hand in glove with the police; nothing is hid from them. . . . They go to bed at dawn, and make their report to the police later in the day.

Though linkmen were associated with authority in Paris, they were, according to William Sidney, akin to thieves in London. To engage a torchbearer there, he insisted, "was an undertaking attended with considerable risk: far more often than not these 'servants of the public' were hand in glove with footpads and highwaymen, and would rarely think twice on receiving a signal from such accomplices of extinguishing the link and slipping away, leaving the terrified fare or fares, as the case may be, to their tender mercies."

Still, the linkmen and linkboys did a brisk business among the well-to-do, who dined out and attended performances and plays. In earlier times, theatrical productions had been staged during the afternoon in the open air or in theaters with large windows or open roofs, and sounds often had to suggest changes in daylight: a cock's crow stood for sunrise, an owl's hoot for night. Now, in evening performances in enclosed theaters, the control of darkness onstage could hide some of the ropes and supports, as well as changes of scenery. Artificial light suggested natural light, and also emotion. In sixteenth-century Italy, it was "a custom, both in ancient and modern times, to light bonfires and torches in the streets, on the housetops, and on towers as a sign of joy; and hence arises this theatrical convention — the imitating of such festive occasions. The lights are put there for no other purpose but to imitate . . . this mood of gaiety."

The candles and lanterns employed as footlights and spotlights, and the *bozze* — candles backed by burnished metal disks and fronted by glass globes filled with tinted water — allowed for a variety of effects. But their presence also meant the illusion had to be ruptured now and again. "Until he himself was snuffed out by the universal employment of gas," wrote one theater historian, "the candlesnuffer had perforce to obtrude himself in the midst of the traffic of the scene to fulfil his humble office. Guttering tallow dips called for immediate attention. . . . When the stage lights began to flare or flicker out the gods commonly set up a cry of 'Snuffers! Snuffers!'"

Sometimes the night city itself could seem like a vast public interior walled in by the dark of the countryside beyond. On a clear evening in eighteenth-century Vienna, one observer noted, "These beautiful lights are laid out so prettily that if one looks down a straight lane . . . it is like seeing a splendid theater or a most gracefully illuminated stage."

Of course, the longer night hours weren't for everyone. The advantages of streetlights went mostly to the young and the wealthy. Many ordinary workers, obligated to rise with the sun, were unable to truly enjoy the extension of the day that streetlights brought. "Night falls," Mercier wrote, "and, while scene-shifters set to work at the playhouses, swarms of other workmen, carpenters, masons and the like make their way towards the poorer quarters. They leave white footprints from the plaster on their shoes, a trail that any eye can follow. They are off home, and to bed, at the hour which finds elegant ladies sitting down to their dressing-tables to prepare for the business of the night." And although ordinary citizens were no longer required to supply sill lights, they felt resentment still, since they were taxed for them.

In time, those tax dollars paid for improved lighting. Artist Jan van der Heyden developed streetlights for the city of Amsterdam in which currents of air washed over the interior glass of the lanterns and kept soot from accumulating. By the mid-eighteenth century, the simple lanterns on the streets of Paris that hung from cables strung across the way were replaced with *réverbères*, oil lanterns with double wicks and two reflectors to augment the brightness of the flame: one reflector above the flame to direct the light downward, another concave reflector beside the flame to direct the light outward. "In the old days, eight thousand lanterns, their candles askew, or guttering, or blown out by the wind, nightly adorned the city, a feeble and wavering illumination broken by patches of treacherously shifting darkness," wrote Mercier. "Now twelve hundred of these oil-lamps suffice, and the light they throw is steady, clear and lasting." But Mercier also claimed that even lamps and linkmen couldn't guarantee enough light in Paris:

> I have known fogs so thick that you could not see the flame in [the] lamps; so thick that the coachmen have had to get down from their boxes and feel their way along the walls. Passers-by, unwilling and unwitting, collided in the tenebrous streets; and you marched in at your neighbour's door under the impression that it was your own. . . . One year the fogs were so dense, that a new expedient was tried; which was, to engage blind men, pensioners, as guides . . . for they know Paris better than those who have made our maps. . . . You took hold of the skirt of the blind man's coat, and off he started, stepping firmly, while you more dubiously followed, towards your destination.

Perhaps no city had a more complex relationship with street lighting than Paris. In the latter part of the eighteenth century, lantern smashing — once simply a rogue's pastime — became not only a symbol of defiance but also a strategy in the rebellion against the state. "The darkness that spread as lanterns were smashed created an area in which government forces could not operate," notes historian Wolfgang Schivelbush.

"Lantern smashing erected a wall of darkness, so to speak." Or returned the streets to their old dark.

And streetlights took on even greater significance in the days after the storming of the Bastille in July 1789. Before the revolutionists adopted the guillotine for retribution — "the clanking of its huge axe, rising and falling there, in horrid systole-diastole" — they chose not signposts or trees on which to hang French officials, but the lanterns. "In the summer of 1789, the meaning of the French verb *lanterner* changed," notes Schivelbush. "Originally, this word meant 'to do nothing' or 'to waste one's time.' At the beginning of the Revolution, it meant 'to hang a man from a lantern.'"

Charles Dickens suggested that "the gaunt scarecrows of that region should have watched the lamplighter, in their idleness and hunger, so long as to conceive the idea of improving upon his method, and hauling up men by those ropes and pulleys, to flare upon the darkness of their condition." Lantern cables were meant to carry the weight of one small light. During the hangings, "not infrequently, the hapless offender had to be taken to four or even six *réverbères* before a rope strong enough to survive this treatment was found." Hangings were not confined to the lanterns strung across the streets on cables. Those in the city's squares, which could not be spanned by a rope, were affixed to the walls, and it was in such a square that Joseph-François Foulon de Doué, controller general of finances under Louis XVI, was hung. Foulon — who had suggested that if the people were hungry, they could browse on grass — was "whirled across the Place de Grève, to the 'Lanterne' . . . pleading bitterly for life, — to the deaf winds. Only with the third rope (for two ropes broke, and the quavering voice still pleaded) can he be so much as got hanged! His Body is dragged through the streets; his Head goes aloft on a pike, the mouth filled with grass: amid sounds . . . from a grass-eating people."

The verb *lanterner* must have meant little to those in the French countryside, who endured in darkness the same hunger and want as the poor of Paris. Once the night had been the same for all; now light began to separate more fully country from city. Little by little, the city night began to influence the rhythms of its day. The privileged and wealthy, who had always been profligate with light — the more their parties and dances were brilliantly illuminated, the greater seemed their position and power — now habitually rose late in the day, so that rising late, too, became a mark of prestige. One of their contemporaries complained that the courtiers altered "the order of nature by making the day into night and the night into day, namely when they stay awake in order to indulge in their entertainments, though other people sleep: afterwards to restore the vigor lost by their sensual pleasure they sleep while other people are awake and attend to their business." As more people stayed up later at night, the hours of the market shifted: merchants' stalls in Paris, which had previously opened at four in the morning, "now [opened] hardly at seven o clock," and shops stayed open after daylight began to fail.

The cities began to develop their own seasonal rhythms as well. In the countryside, as the days grew shorter and colder, everything began to draw in: the birds scratched at bark and scraped at snow, the sheep huddled in their folds, and people lived confined to the one or two rooms warmed by their fires. Meanwhile, in the city the streetscape seemed to grow livelier, as the wealthy returned from their summer refuges, and the night seasons of the opera, theater, and ballet began. In winter, the light and warmth of cafés and taverns seemed particularly inviting. By the twentieth century, one observer would claim: "The city lives at cross-purposes with nature: cold not heat brings it to life. . . . It is during the fall and winter that the sense of renewal is at its height, for as one place dies another comes to life."

The greater the hours of illumination, the more the city at night worked its way into the human imagination, until the illuminated city and the glamour and liveliness of its night came to define almost completely what it meant to be urban and urbane, and any metropolis possessing less than a brilliant, vibrant night was deemed provincial. Later, in the twentieth century, Elizabeth Hardwick would write that Boston was

> not a small New York, as they say a child is not a small adult but is, rather, a specially organized small creature. . . . In Boston there is an utter absence of the wild electric beauty of New York, of the marvelous excited rush of people in taxicabs at twilight, of the great Avenues and Streets, the restaurants, theaters, bars, hotels, delicatessens, shops. In Boston the night comes down with an incredibly heavy, small-town finality. The cows come home; the chickens go to roost; the meadow is dark.

3

LANTERNS AT SEA

ALTHOUGH EIGHTEENTH-CENTURY CITIES had begun to emerge from their ancestral night, the world's oceans remained couched in it. Ships under sail or at anchor might display a lantern on deck, but lamps carried the old danger of fire to the cramped holds of wooden sailing vessels. To avoid disaster at sea, merchantmen often dressed and ate in the dark — for them, wrote Herman Melville, oil was "more scarce than the milk of queens." Emigrant ships prohibited lamps below deck, although some travelers were allowed to carry enclosed lanterns. Slaves were denied any light at all.

If something blazed on the seas, it was likely a whaling ship in the hours after a catch — its try-pots boiling and smoke fogging the rigging, while almost all hands flensed the blubber from the whale and minced it into portions to feed the vats. "The oil is hissing in the trypots," wrote J. Ross Browne.

Half a dozen of the crew are sitting on the windlass, their rough weather-beaten faces shining in the red glare of the fires, all clothed in greasy duck. . . . The cooper and one of the mates are raking up the fires with long bars of wood or iron. The decks, bulwarks, railing, try-works, and windlass are covered with oil and slime of black-skin, glistering with the red glare of the try-works. Slowly and doggedly the vessel is pitching her way through the rough seas, looking as if enveloped in flames.

During the eighteenth century, hundreds of whaling vessels sailed the seas in search of their quarry, for though many people still made or bought tallow candles, and those in continental Europe often fed their lamps with colza (rapeseed) oil, whale oil — cheap and abundant — fueled much of the growing brilliance of the domestic and municipal night. Common whale oil was also called "train oil," from the Old High German word *trahan*, meaning "drop" or "tear," because, it's said, the oil was originally pressed bit by bit from the blubber. It ranged widely in quality and price. The most expensive pale oil burned clear and clean, while cheaper brown oil — usually rendered from old blubber — smoked readily and stank of old fish.

The whaling trade had evolved over thousands of years, having begun with the harvesting of stranded whales. Whales have always inexplicably beached themselves, and once out of their element, they can't survive for long. Exposed fully to the sun, their skin burns, and they are crushed by their own weight. "When the flesh has moldered away, the skeletons are left, which the inhabitants of these shores use for building their houses," explains a perhaps fanciful account from the time of Alexander the Great. "The large bones at the side form the beams of their houses, the smaller ones the lathes. From the jawbones they make doors." Whether or not people made houses of whalebone, those living along protected bays throughout the world cut up the carcasses for food and rendered the fat to use for fuel and lubricants. What remained after the harvest they let wash away with the tide.

As the demand for blubber and whalebone grew, men took to their boats and forced the whales toward shore. "When they come within our harbors, boats surround them," one early New Englander wrote. "They are as easily driven to the shore as cattle or sheep are driven on land. The tide leaves them, and they are easily killed." This method of herding whales was never as effective as harpooning them, a practice that the Basques along the Bay of Biscay undertook as early as the tenth century. In time, harpooning became the favored method for hunting whales throughout the world.

Early whalers eagerly took any whale they encountered, though by the eighteenth century they were on the hunt for *Eubalaena glacialis*, the North Atlantic right whale, and *Eubalaena australis*, the southern right whale — so named because they were the right whales to hunt. Their blubber yielded good-quality oil, they were slow moving enough to be harpooned, and they floated after death — unlike the blue whale, *Balaenoptera musculus*, which swam too fast for harpooners to reliably capture them and, if struck, invariably sank upon dying. The right whale — black, with white patches on its belly and callosities mottling its face — possesses a massive broad back; the Japanese call it *semi kujira*, meaning "beautiful-backed whale." It can reach sixty feet in length and weigh more than eighty tons. "The respiratory canal is over 12 inches in diameter," observed William Davis, "through which the rush of air is as noisy as the exhaust-pipe of a thousand-horse-power steam engine; and when the fatal wound is given, torrents of clotted blood are sputtered into the air over the nauseated hunters. . . . [Yet] the right whale has an eye scarcely larger than a cow's, and an ear that would scarcely admit a knitting needle."

As for the blubber, or blanket, of the whale, Davis suggested that it could "carpet a room 22 yards long and 9 yards wide, averaging half a yard in thickness." He continued: "The lips and throat . . . should yield 60 barrels of oil, and, with the supporting jaw-bones, will weigh as much as twenty-five oxen of 1,000 pounds each. Attached to the throat by a broad base is the enormous tongue, the size of which can be better conceived by the fact that 25 barrels of oil have been taken by one. Such a tongue would equal in weight ten oxen."

Old Norse legend has it that for all its bulk, the right whale "subsists wholly on mist and rain and whatever falls into the sea from the air above." It could seem so, as it plows through the ocean's surface, taking in gray-green water swarming with krill, plankton, and schooling fish. The right whale is a filter feeder: it strains seawater through the hairy fringes of its comblike baleen plates — "that wondrous Venetian blind," Melville called them — which every day capture more than a ton of small and microscopic sea life. The plates — each up to seven feet long — hang from its upper jaw, and there are more than two hundred pairs in the whale's mouth. Baleen is composed of sturdy, flexible keratin — the same substance as our fingernails — and in the eighteenth century, baleen brought in good money for whalers because it was perfect for making the products of the day: umbrella handles, buggy whips, fishing rod tips, carriage springs, tongue scrapers, shoehorns, boot shanks, divining rods, policemen's clubs, and corset stays. Yet the blubber was a greater prize. The rendered fat from one right whale — after being strained and bleached on shore — could yield more than 1,800 gallons of oil: 60 barrels, each containing 31½ gallons.

"It is as if from the open field a brick-kiln were transported to her planks," wrote Herman Melville of a ship's tryworks.

The timbers beneath are of a peculiar strength, fitted to sustain the weight of an almost solid mass of brick and mortar, some ten feet by eight square, and five in height. . . . On the flanks it is cased with wood, and at top completely covered by a large, sloping, battened hatchway. Removing this hatch we expose the great try-pots, two in number, and each of several barrels' capacity. . . . Sometimes they are polished with soapstone and sand, till they shine within like silver punch-bowls.

In earlier times, the trying out had been done on shore, but as whaling trips grew longer due to increased demand for oil and whalebone — and the increased scarcity of whales — the blubber, especially in warm weather, would begin to spoil over time. Residents of whaling ports, who had suffered the stench and smoke of the tryworks, may have welcomed the move to shipboard rendering, for there was nothing refined about the business. The first fire of a voyage would be lit with wood, but "the unmelted skin of the whale made a wonderful fuel, and the whale was therefore cooked in a fire of his own kindling." The green hands had to get used to the smell — "like the left wing of the day of judgment" — for in flush times the try-pots could boil for a week, and the hands minced unceasingly for their 1/150th of a share. At any moment, the sails and rigging might catch fire while they were trying out, and the ship would burn to the waterline. If the seas were rough, the boiling oil could splash and scald them; they could be crushed by slabs of blubber as they stripped it from the whale, which hung from chains off the side of the ship. They might slash themselves with the sharp blades of their tools or slip on decks slick with oil and blood as they cut the blubber into slabs called blanket pieces. These were cut into horse pieces — so named because they were cut on sawhorses — which were then cut partway through into thin slices, or "leaves," so that the slab remained intact on one side, looking very much like the pages of a book. "'Bible leaves! Bible leaves!' This is the invariable cry from the mates to the mincer," wrote Melville. "It enjoins him to be careful, and cut his work into as thin slices as possible, inasmuch as by doing so the business of boiling out the oil is much accelerated and its quantity considerably increased, besides perhaps improving it in quality."

After the men finished rendering the blubber, they dumped what remained of the carcass back into the sea amid a frenzy of sharks, then scoured the ship with lye leached from the cinders and ashes of the fires. No matter the scrubbing, they never got rid of the smoky stench, which seeped into the wooden deck, the canvas sails, their clothes, and their own pores. It was said that sailing vessels downwind of a whaling ship could smell it from miles away. The only "clean ship" was one that returned to port without any oil in its hold.

Upon their return to port, the hands might have nothing to show for their time at sea, as they'd borrowed against their share of the profits, so they had no choice but to ship back out again, into the drift ice and doldrums and winter storms, the fevers and scurvy, the salt pork and hardtack. Is it any wonder that Melville imagined that the men squandered what oil they could? "There they lay," he wrote of the *Pequod*'s crew, "in their triangular oaken vaults, each mariner a chiselled muteness; a score of lamps flashing upon his hooded eyes. . . . See with what entire freedom the whaleman takes his handful of lamps — often but old bottles and vials, though — to the copper cooler at the try-works, and replenishes them there, as mugs of ale at a vat. He burns, too, the purest of oil . . . sweet as early grass butter in April."

By the mid-eighteenth century, as city streets and homes grew brighter and the demand for oil (and whalebone) continued to grow, the number of ships pursuing and capturing whales increased. During the years just before the American Revolution, more than 360 whaling vessels sailed from New England and New York alone, and the industry had not yet reached its height. But after centuries of persistent slaughter, the right whale had grown scarce in its known grounds, and of necessity whalers undertook even longer trips to farther, deeper waters in pursuit of any kind of whale that might prove to be profitable. The English and Dutch headed north in search of the polar whale — or whalebone whale, as the seamen called it — known now as the bowhead whale, *Balaena mysticetus*, also a filter feeder. The New England fleet sailed to grounds off Newfoundland, along the coast of Labrador, to the west of Greenland, and farther, seeking not only the right whale but also the sperm whale, *Physeter macrocephalus* — also known as the cachalot. This whale travels the world's oceans, heading north in summer and into tropical waters in winter in pursuit of squid. The fleet followed it to the Arctic and as far as the South Pacific.

Although the sperm whale isn't a filter feeder, and thus has no baleen, and one whale might yield only twenty-five to forty-five barrels of oil (far less than a right whale), the quality of the harvest made it worth the chase. At its best, the oil from a sperm whale burned clear and clean and was almost odorless. But even more valuable was the spermaceti, a waxy substance found in the head which could be made into candles of the highest quality. These candles had a high melting point and gave off twice the light of candles molded from tallow. And the flame did not smell foul — a quality dearly valued in an age of sputtering, stinking tallow candles and dim, finicky grease lamps, which also stank. Spermaceti candles were probably first molded around the mid-eighteenth century; it's likely that Benjamin Franklin was referring to them when he wrote of "a new kind of Candles very convenient to read by. . . . They afford a clear white Light; may be held in the Hand, even in hot Weather, without softening. . . . They last much longer, and need little or no Snuffing."

The sperm whale can grow to more than sixty feet, weigh more than sixty tons, and possess a blanket of blubber almost a foot thick. But its greatest feature is its massive, scarred, and battered head, flecked with the sucker marks of squid. According to Herman Melville,

> In the great Sperm Whale, this high and mighty god-like dignity inherent in the brow is so immensely amplified, that gazing on it, in that full front view, you feel the Deity and the dread powers more forcibly than in beholding any other object in living nature. For you see no one point precisely; not one distinct feature is revealed; no nose, eyes, ears, or mouth; no face; he has none, proper; nothing but that one broad firmament of a forehead, pleated with riddles; dumbly lowering with the doom of boats, and ships, and men.

That head houses a brain of about eighteen pounds — the largest on earth — and contains two large cavities, known to whalers as the "case" and the "junk." The case, at the top of the head, is full of a mixture of oil and spermaceti — also called "head matter" — an almost-clear amber or rose-tinted waxy liquid that whalers hauled out of the carcass with buckets. Once out of the whale and exposed to cold air, the head matter crystallized and hardened to a pure white mass, which was stored in barrels for the rest of the voyage. There could be up to five hundred gallons of it in an average sperm whale, nine hundred in a large bull.

The junk, located in the lower half of the forehead, contains a spongy material impregnated with sperm oil. The oil squeezed from the junk made the finest lamp oil. Additionally, whalers harvested oil from the blubber of the sperm whale. The price of oil always depended on supply and demand, as well as on the quality of the oil, which varied from whale to whale even within a given species. But sperm oil always fetched a price three to five times that of common whale oil. In 1837, when the annual sperm oil yield of the American fleet was more than 5 million gallons, it sold for $1.25 per gallon. The price peaked in the 1860s at $2.55 per gallon.

Unlike tallow, spermaceti couldn't be dipped or molded into candles by a housewife in her kitchen, for the complex process of making spermaceti candles took almost an entire year to complete. After the spermaceti arrived in port, it was brought to the candle works, where the candlemakers boiled it to filter out impurities and then stored it until the cold weather, when it would fully congeal. On a mild winter day, when the spermaceti softened a bit, they shoveled it into woolen bags and pressed it between the wooden leaves of a large screw press. The oil they squeezed from the spermaceti then was called "winter strained sperm oil" — clear and clean — and they sold it as lamp oil, which commanded the highest price. They stored the remains until spring, when they heated it again to filter out more impurities, then cooled it, molded it into cakes, and shaved it into small pieces before they pressed it again — this time in cotton bags and under greater pressure — to produce "spring strained sperm oil," which was a lower-quality oil. What remained in the bags they pressed a third time to make "tight pressed oil" or "summer oil." The remaining solid after these three pressings was almost pure spermaceti — waxy, brownish or yellowish in color, and streaked with gray. They stored it for the summer, then heated it again, this time with potash to bleach and clarify it — clear as spring water, it was said — before molding it into candles that would fetch twice the price of those made of tallow. Spermaceti candles had no comparison, except perhaps those made from beeswax, and like beeswax candles, they would always remain the province of the well-to-do. So steady and clear was their light that the brightness of the flame of a pure spermaceti candle that was seven-eighths of an inch in diameter and weighed one-sixth of a pound would eventually become a standard of measure for luminous intensity — one candlepower — against which the light of other candles, all lamps, and even the first electric lights would be measured.

The desire and demand for spermaceti and sperm oil would drive the whaling trade well into the nineteenth century. The size of the fleet reached its peak around 1846, when more than seven hundred vessels sailed out of twenty major American ports and a host of smaller ones, and several hundred vessels from other countries also roamed the whaling grounds. The oil and spermaceti brought to port that year were valued at $8 million. Melville himself posed the question as to "whether Leviathan can long endure so wide a chase, and so remorseless a havoc; whether he must not at last be exterminated from the waters, and the last whale, like the last man, smoke his last pipe, and then himself evaporate in the final puff." There had been an estimated 1.1 million sperm whales in the world's oceans before the hunt for them began in earnest. How many remained in the mid-nineteenth century isn't known, but it's thought that today the somewhat recovered population stands at about 360,000.

Although sperm whales may have been the prized catch, whaling ships continued to take whatever whales they could find. In 1851 more than 10 million gallons of common whale oil — bringing about 45 cents per gallon — also arrived in

American ports. The millions of gallons of whale oil and sperm oil circulating the globe meant that light was more readily available to many people — especially those in cities — than it had been in the past. People began to light more household lamps in the evening and leave them lit for a longer time. More light, yes, and unlike the old local oils and tallows, light at a far remove from its grimy source, so people for the first time could distance themselves from the whole endeavor of light's production. Melville's Ishmael said of himself and his companions, "They think that at best, our vocation amounts to a butchering sort of business; and that when actively engaged therein, we are surrounded by all manner of defilements. Butchers, we are, that is true. . . . But, though the world scouts at us whale hunters, yet does it unwittingly pay us the profoundest homage, yea, an all-abounding adoration! for almost all the tapers, lamps, and candles that burn round the globe, burn, as before so many shrines, to our glory!"

Eighteenth-century whalers suffered their own particular perils, but the seas were dangerous for all sailors. Most navigation tools were rudimentary and the charts imprecise. Once night fell or weather closed in the coast, mariners had few lights to help them steer clear of sandbars, stone reefs, or the debris of old wrecks. What lights there were — often no more than open coal or wood fires on the headlands and burning baskets of pitch or oakum atop long poles — were of limited help, for they barely penetrated the fog and the dark and didn't always stand up to the prevailing winds and storms that battered them. The work of keeping the light alive could be unceasing: on a windy night, an open fire could consume a ton of coal or countless logs. The endless demand for fuel for coastal fires was one of the primary reasons for the deforestation of the island of Anholt off the coast of Denmark.

The smoke-clouded lanterns of the few lighthouses were no better. Their flames — sometimes open to the elements, at best enclosed in glass or horn and magnified with reflectors or convex lenses — were small and unsteady. The lamps, often possessing multiple wicks, had to be constantly snuffed, guarded, fanned, and fed. They were hard to light in the cold, and the keepers — ill supplied, isolated, and miserably paid, themselves barely protected from the wind and rain — might need to place hot coals near the lanterns to prevent the oil from congealing. Despite the best efforts of the keepers, along the British coast alone — the best-lit coast in the world in the early eighteenth century — more than five hundred ships foundered every year.

At times it was lights themselves that sunk ships, for well-intentioned beacons could be deceiving. Almost all of them in the eighteenth century were fixed — there was no system of flashing lights to help distinguish one lighthouse from another as there would be in later times — and although a fixed light could help orient those who knew the waters, it was of little help to someone unsure of his bearings. A ship approaching land after a long, wind-tossed voyage could be far enough off course that the navigator might mistake the light he encountered for a different one farther along the coast. Or a light, being precarious, would go out, and the navigator might find no light where he expected one to be. It was also true that a terrestrial light might appear to be celestial. Pliny the Elder, speaking of seamarks in Roman times, wrote, "The only danger is, that when these fires are thus kept burning without intermission, they may be mistaken for stars, the flames having very much that appearance at a distance."

Lights could also be intentionally deceiving: wreckers intent on stealing washed-up cargo sometimes set a lantern on a dark headland hoping it would be taken for a true seamark, though their usual method, wrote lighthouse historian D. Alan

Stevenson, "was to drive an ass bearing 2 lanterns along the shore, to represent a vessel in motion and so lure a ship to destruction among the near rocks and shoals." Wreckers weren't the occasional wayward souls. Historian Bella Bathurst notes, "Many coastal villages staked their livelihoods on the exotic plunder to be found in dead and dying ships; the wreckers saw their lootings as a perk of nautical life, and bitterly resented any attempt to interfere. . . . The wreckers were furious at the prospect of a safer sea."

It was said that the open flame of the first known lighthouse, the Pharos, could be seen a hundred miles away. Although that is certainly an exaggeration, the Pharos was an impressive structure. Built for the port of Alexandria in the third century B.C., its light — which was intensified and projected by a curved mirror or polished metal disk — was housed in the cupola of a rectangular marble structure that rose about four hundred feet above the low-lying Egyptian shore. At the time, only the pyramids stood taller. By comparison, eighteenth-century shore lights were far more modest, and on a clear night a well-maintained beacon might be seen five, six, maybe seven miles away, which was far short of some of the worst ocean perils. For instance, the rocks of the Eddystone reef, which lie nine miles off the south coast of England, extend for half a mile, and nearly all of them are submerged, the most prominent rising only three feet above water during the highest tides. According to Bathurst,

> The rust-colored gneiss is as resilient as diamonds, and the currents that surround it send up abrupt spouts of water even on the calmest days. It is thought of as a bad-tempered place, full of sulks and strange moods, and by the sixteenth century its reputation for destruction had already spread well beyond

Cornwall. . . . Merchant captains were so alarmed by the prospect of being wrecked on the Eddystone that they often ran themselves aground on the Channel Islands or the northern French coast trying to avoid it.

It was at Eddystone, on rock fully exposed to the sea, that the first offshore light, engineered and built by Henry Winstanley, was completed in 1698. Winstanley secured the structure by driving twelve iron rods into the highest rock on the reef. He then surrounded the rods with stone. Glaziers, smiths, masons, and carpenters made trips from Plymouth almost daily when the weather held. They moved tons of material from their boats to the rock even in rough seas and accomplished their work as the tides rose and fell around them. D. Alan Stevenson wrote:

> At midsummer the party decided to lodge in the tower, hoping to save the time and labour spent in passage between Plymouth and the reef. But during the first night a storm of exceptional severity for the season arose unexpectedly and no boat could approach to take them off. With little shelter . . . they were marooned in the roofless tower for 11 days [and finally] got ashore in a half-drowned condition. When the weather improved, undeterred by the unhappy experience, they returned to complete the lighthouse and lighted it on the 14th November. . . . In the following months the waves over-topped the lantern and Winstanley saw that he must raise it.

The next year, Winstanley built an almost entirely new structure, forty feet taller than the first. His second light lasted three years before severe winter weather damaged it. When Winstanley returned to the rock again to oversee repairs, he, his workmen, and the keeper were caught on the reef during one of the fiercest storms ever recorded along that coast. After the weather cleared, there was no sign of any man, and all that was left of the light were a few twisted pieces of metal — remnants of the rods that had tied the tower to the rock.

The third tower at Eddystone — a timber sheath packed with stone, built by John Rudyard — stood for fifty years until the wooden lantern that housed the flame caught fire. The light from the conflagration, seen from the English shore, reached farther than the beacon ever had. According to Stevenson,

> Quickly the fire got a grip of the tower, the flames extended downwards over their heads and drove the men from room to room until they found shelter in a cleft in the Rock . . . while burning embers and red-hot bolts rained down. . . . One of the lightkeepers . . . declared that when looking upward during their descent of the burning tower, a quantity of molten lead had fallen into his mouth and down his throat. He experienced no pain and a physician who examined him did not believe his tale, but he died twelve days later. . . . The dreadful experience at the Eddystone so terrified another of the lightkeepers that on reaching land he ran off and was not heard of again.

Yet a fourth tower was planned for the reef, this time designed by engineer John Smeaton, who based his plans for it on the shape of an English oak tree, believing that a flared base would give the tower greater stability. The innovative design would be the model for lighthouse construction for more than a century. Smeaton built his light entirely of stone, using granite for the foundation and exterior and softer Portland stone for the interior. Masons in the coastal city of Plymouth began cutting one-ton stones in August 1756, and the following summer they began to build the light. Stevenson wrote:

> Fenders fixed on the east side of the Rock prevented boats from fretting against it. Shears and a windlass were fixed for raising the stones directly from a boat and tested by hoisting above the

Rock a heavy longboat complete with crew. . . . Sunday 12th
June saw the first stone, weighing 2 tons fitted in position and
bedded with mortar. . . . Next day the masons set the other 3
stones of the 1st course. On the 15th a heavy swell carried away
5 of the 13 stones . . . but the masons at Plymouth working day
and night, cut duplicates in 2 days.

They sometimes worked into the summer nights, seeing only
by the flickerings of lighted links, or torches, and still it took
more than three years to finish the tower, which weighed more
than a thousand tons and stood eighty feet above the rocks.
First lit in October 1759, the light shining from Smeaton's
tower was no different, or stronger, than that in the previous
Eddystone lights: a chandelier of twenty-four candles (each
about the size of a contemporary dinner-table candle), which
the keeper lowered every half-hour for snuffing, then raised
again into place. If the glass was clean and the light well
snuffed, it could be seen for seven miles: "very strong and
bright to the naked eye, much like a star of the fourth magnitude." It stood on the reef for 120 years.

Traffic on the seas increased markedly during the eighteenth
century, and the story of the Eddystone light illustrates the
lengths to which people would go to achieve even a small glimmer of illumination. They had no hope for more than that,
really. In spite of the widespread slaughter of whales, the stink
of try-pots, and the complex process of making spermaceti
candles, eighteenth-century light wasn't appreciably brighter
than what could be had in Roman times, for lamp technology
had hardly changed, in part because not even the scientists
of the time understood the nature of the flame they were gazing into at night. What would eventually bring about the first

measurable increase in the brightness of lamps occurred a world away from the oil-slicked decks of whaling ships, in the laboratories of Europe.

At the time of the French and American revolutions, scientists adhered to the belief that all matter contained phlogiston, a flammable substance that was imparted to the air during combustion. "So long as the air can receive this substance from the combustible matter so long the body will continue burning," noted Professor Samuel Williams, who lectured at Harvard at the time.

> As soon as the Air is saturated and can receive no more of the Phlogiston, the combustion must cease for no more Phlogiston can escape or be thrown out from the burning body. And therefore when fresh air is admitted to receive Phlogiston, the combustion will again take place. — And hence are derived the phrases of *phlogisticated* and *dephlogisticated* air. By phlogisticated air is intended air which is charged or loaded with Phlogiston, and by dephlogisticated air is meant Air which is free from Phlogiston; or which does not contain this principle or element of inflammability.

Quite a few scientists experimented with combustion in the last quarter of the eighteenth century, most notably Joseph Priestley in England and Antoine Lavoisier in France. Eventually, Priestley identified oxygen in air, although he continued to hold fast to the phlogiston theory. It was Lavoisier, working in Paris, who built on Priestley's understanding of oxygen and concluded that rather than imparting a substance to the air, burning materials were fueled by oxygen in the air.

François-Pierre Ami Argand, a Swiss scientist who worked briefly in Lavoisier's laboratory, made use of his and Priestley's findings to create the first significant improvement in the lamp. The most essential component of Argand's design was a tubular wick, which he fed between two metal cylinders. Openings at the base of the cylinders allowed air to reach the flame from both inside and outside the wick. The increased oxygen created a more robust flame than in previous lamps, and it also burned at a higher temperature, making for a cleaner fire in which the carbon particles were almost completely consumed. An Argand lamp produced very little soot and smoke, and there was little need for snuffing. Later, Argand enclosed the wick in a chimney — perforated metal, then glass — which not only protected the light but also created an updraft that increased airflow to the flame. He also designed a mechanism for raising and lowering the wick. According to some accounts, his lamp shone more brightly than six tallow candles. Others claimed that if it was fed by spermaceti oil, it produced about ten times the illumination of a customary lamp, and the flame — rather than being the usual orange — was "very white, lively and almost dazzling, far better than the light of any lamp proposed before."

This light born of experiment, of the investigations of a handful of men in private quarters, seemed so immediately bright that to some it was more than the human eye could bear. One account suggests that "as the light emitted by [these lamps] is frequently too vivid for weak or irritable eyes, we would recommend the use of a small screen, which should be proportionate to the disk of the flame, and be placed, at one side of the light, in order to shade it from the reader's eye, without excluding its effect from others, or darkening the room." And, after so many centuries of dreaming of more light, people did shield the flame, with mica, horn, and decorative glass. These were the first lampshades.

The Argand lamp had its challenges. Though efficient, the

large wick and increased oxygen required much more oil than previous lamps, which not only made the lamp costly to run but also meant that Argand couldn't count on capillary action alone to feed the flame, since the viscous animal and vegetable oils of the time rose so slowly up the wick. To solve this problem, Argand designed an oil reservoir adjacent to and higher than the burner, which used gravity to feed fuel to the lamp, but the reservoir partially obscured the light and cast a shadow.

"Being 'the thing,' the Argand or Quinquet lamps [as they were known in France] were usually made up in bronze, silver, porcelain, crystal, and other expensive materials that kept them well out of reach of the ordinary purse," observes historian Marshall Davidson. And it wasn't just the cost of the lamps that kept those of meager means from buying them; the quantity of oil required stopped them as well. Brilliance still came at a price, and they knew it. "The modest versions that Yankee tinsmiths were advertising as early as 1789 did not win any broad popularity," notes Davidson. "Absurd as it sounds they gave too much light. That is to say, it was impracticable to make them so small that they had no greater flame than that of a single candle and . . . anything that burned more oil, proportionately, whatever its brilliance and efficiency, was uneconomical for ordinary domestic purposes."

For mariners, the Argand lamp was invaluable. A lighthouse equipped with one magnified by a parabolic reflector not only gave many times the light of the old lighthouse lamps, but the light proved steadier and more dependable. The adoption of the Argand lamp for seamarks, along with an increase in lighthouse construction, meant, according to Stevenson, that "the single most powerful light of 1819 exceeded the combined powers of all the navigation lights of 1780." And perhaps the greatest innovation, one used even now, was still to come.

In 1822 French physicist Augustin-Jean Fresnel designed a hive of light. His Fresnel lens — a lamp comprising concentric wicks set in bull's-eye glass and surrounded by rings of glass prisms — bent and concentrated light into a bright, narrow beam. The largest of his lenses, meant to aid ships along the most treacherous and fogbound coasts, was built of a thousand prisms and stood more than ten feet high. When placed one hundred feet or so above sea level — high enough to compensate for the curvature of the earth — its beam could be seen for twenty miles. Fresnel produced his lens in six different sizes; the smallest, a sixth-order lens used in harbors and bays, was a mere twelve inches in diameter and stood eighteen inches high.

Throughout the nineteenth century, in addition to installing Fresnel lenses and replacing old oil lamps with more dependable electric lights or gaslights, lighthouses would begin to adopt a system of flashing lights to distinguish one seamark from the next. Mariners unfamiliar with the coast could get their bearings even when daymarks — the painted patterns on lighthouse towers — disappeared with the sun. And lightships, light buoys, and sound signals such as whistles, bells, and foghorns frequently marked the more treacherous shoals.

Still, shipwrecks were a given well into the twentieth century. In the early 1920s, there were twelve working Coast Guard stations along fifty miles of the south shore of Cape Cod, and lantern-carrying surfmen patrolled the shores, scanning the waters for ships in distress. "Every night they go; every night of the year the eastern beaches see the coming and going of the wardens of Cape Cod. Winter and summer they pass and repass, now through the midnight sleet and fury of a great northeaster, now through August quiet . . . the beach traced and retraced with footprints that vanish in the distances," observed Henry Beston, who chronicled life on "the Great Beach of Cape Cod."

> There has just been a great wreck, the fifth this winter and the worst. . . . The big three-masted schooner *Montclair* stranded at Orleans and went to pieces in an hour, drowning five of her crew. . . . Older folk will tell you of the *Jason*, of how she struck near Pamet in a gale of winter rain, and how the breakers flung the solitary survivor on the midnight beach; others will tell of the tragic *Castagna* and the frozen men who were taken off while the snow flurries obscured the February sun. Go about in the cottages, and you may sit in a chair taken from one great wreck and at a table taken from another; the cat purring at your feet may be himself a rescued mariner.

Any mariner of the eighteenth century would have found it impossible to comprehend that one day a marker on the Eddystone reef would emit a light equivalent to 570,000 candles, or that such a light would not be essential to seeing a ship safely past the rocks; that there would come a time when navigators hardly needed to scan the horizon, for they would get their bearings from a prism of information — radar, GPS, and electronic charts. Data would become the new lamp.

4

GASLIGHT

AT THE TURN OF the nineteenth century, most people still saw by the same ancient light as always, though that would change in the decades to come. Not only would brighter, cleaner mineral fuels replace tallow and whale oil, but the story of human light would cease to be that of candles and lamps alone. It would become a story that defied linearity, one composed of inseparable strands of invention and improvement — gaslight, the safety match, electric arc lamps, kerosene, Edison's incandescent bulb, Tesla's alternating current — and as new forms of illumination overtook the old, they competed with one another in ways that stratified society and intensified the separateness of countryside and city, household and industry.

In the first decades of the nineteenth century, gaslight led this transformation, at least for city dwellers and factory workers in England. The gas fuel of the time was a by-product of the distillation of bituminous coal into coke (the "charcoal of coal"), and coke production was well established in England, whose economy had been based on coal for more than a century. The English preferred burning hard, light, porous coke in both their home hearths and industrial furnaces. Unlike bituminous coal, which in its raw state burns with a smoky yellow flame, coke burns with a uniform and intense heat that produces no sparks and little soot or smoke. "It seldom needs the application of the poker — that specific for the *ennui* of Englishmen," noted one writer of the time.

Coke manufacture involved shoveling coal into vessels called retorts, which were set in large ovens and heated — a process that dissipated the tar and gases present in the coal. During the eighteenth century, coke manufacturers captured and sold the tar, which was used for caulking ships, but they released the coal gas into the air and let it go to waste. Although it had long been known that such gas would burn with a luminous flame and scientists had experimented with igniting bladders filled with coal gas and other flammable substances, until the turn of the nineteenth century, no one had developed a practical application for it.

In 1801 French engineer Philippe Lebon gave the first public demonstration of functional gaslight when he displayed, in Paris, what he called the *thermolampe*. This furnace housed a retort that fed distilled flammable gas — likely wood gas — to a condenser, then through a series of pipes to an outlet. Lebon imagined that his thermolampe would be used for both lighting and heating a household: "The inflammable gas is ready to extend everywhere the most sensible heat and softest lights, either joined or separated at our pleasure. In a moment we can make our lights pass from one chamber to another. . . . No sparks, coals or soot will incommode us any longer. Neither can cinders ashes coals or wood, render our apartments black or dirty or require the least care." He outfitted his own home with a thermolampe and sold admission for viewing it in an effort to arouse public interest. Many were curious, few were persuaded, and the thermolampe went no further.

Gaslight found its first sustained application as light alone in British machine shops and cloth factories, where the limits of tallow and whale oil were keenly felt. This was especially true in the winter, when the working day continued long after darkness fell, and the wavering light cast by such illuminants made precision work difficult. To light their workrooms, some large factories needed hundreds, even thousands, of tallow candles or whale oil lamps. Each required individual attention — lighting, snuffing, replacing, filling, cleaning — never mind the stink, the irritating smoke, and the heat. In addition, any simple accident could mean disaster. Some owners of large factories so feared a conflagration that they kept their own fire engines on hand. Such light was costly, too. According to historian M. E. Falkus,

> All factories . . . used considerable quantities of oil and tallow in winter months. In 1806, one of the largest of Manchester's spinning factories, McConnel & Kennedy, burned candles for at least eight hours on the shortest days and averaged four hours lighting a day for six months of the year. . . . The annual cost of lighting McConnel & Kennedy's factory in 1806 was about £750. This firm burned an average 1,500 candles each night for 25 weeks in the year and consumed more than 15,000 lbs. of tallow.

William Murdoch, chief engineer at Boulton and Watt — one of most prominent firms in England and builder of the first steam locomotive — experimented with coal gas at the same time Lebon was developing his thermolampe. Although others were also considering how to use coal gas, Murdoch achieved the first real success. His system differed from Lebon's only in its scale: he fitted retorts with pipes that carried distilled gas to huge reservoirs or storage tanks, called gasometers, and fitted the gasometers with outflow pipes, which could send gas, when needed, through mains and then smaller pipes to outlets.

Murdoch lit his own cottage for his initial experiment, and then in 1802 he built a larger system for Boulton and Watt's forge in their Soho, Birmingham, factory. Its success led him to expand the system to include the workshops in Soho. In 1805 he began construction of a gaslight system for the Phillips & Lee cotton mill in Manchester, which he completed several years later:

> It was estimated that more than 900 burners produced light equivalent to 2500 tallow candles burning on average for 2 hours on each working day. The factory contained eleven gasometers, six retorts, and more than two miles of pipes. Total expenditure on the plant was in excess of £5000, the cost of gas was about £600, allowing for depreciation of the equipment and the sale of the coke manufactured as a by-product. . . . The equivalent light produced by tallow candles would have cost an estimated £2000 a year.

These very first gaslight systems probably didn't significantly improve the quality of light in the workrooms. Most observers of the time claimed that one gas burner gave a light three to six times brighter than a common oil lamp, but they had no accurate way of measuring the difference, only a comparison of shadows, which at the time was explained this way:

> Suppose it were required to know how many candles, of a given size, were equal to a patent [Argand] lamp: — place the lamp at one end of the mantle-piece [*sic*], and the candles at the other; hold up the snuffer-tray, a book, or any other object of which the shadow can be received on a sheet of white paper against the opposite wall; the object must be held in a line with the middle of the mantelpiece: the lamp will produce one shadow

and the candles another; when the shadows are equally dark the lights are equal; the darkest shadow will be produced by the strongest light.

To its advantage, a gas flame could be larger than an oil lamp's because it wasn't restricted by the size of the wick, and under ideal circumstances coal gas's combustion was almost complete: it burned with a whiter, clearer flame (in contrast to the reddish orange glow of most simple oil lamps and candles). Yet in the beginning, gaslight was far from perfect. There were few filters for the coal gas, which contained both hydrogen sulfide and carbonic acid, so a foul smell accompanied the light. (Although Murdoch's system for Phillips & Lee filtered the gas through lime, which absorbed the hydrogen sulfide and carbonic acid, this did not entirely purify it.) The gas itself was of uneven quality, its delivery was unreliable, and the equipment was crude. As William O'Dea notes, "The burners were simply iron tubes with holes pierced in them; and apart from the variable and often poor illuminating quality of [the] gas produced . . . the burners quickly corroded and, even when new, over-cooled the flame." Still, the jets didn't require individual attention, and there was nothing to spill or tip. And although gas left a sooty residue, it was cleaner, too.

If gaslight was cleaner, the grime of getting the coal to produce it rivaled that of the hunt for whale oil, as a descent into any British coal mine in the early 1800s would attest. According to a writer of the time,

> Clean and orderly [the miners] coolley [sic] precipitate themselves into a black, smoking, and bottomless-looking crater, where you would think it almost impossible human lungs could play, or blood dance through the heart. At nearly the same moment you see others coming up, as jetty as the object of the

search, drenched and tired. I have stood in a dark night, near the mouth of a pit, lighted by a suspended grate, filled with flaring coals . . . the pit emitting a smoke as dense as the chimney of a steam-engine; the men, with their sooty and grimy faces . . . their sparkling eyes.

Except for that suspended grate at the mouth of the shaft, pitmen would have had almost nothing to see by. They used their candles sparingly, since methane gas — known as firedamp and present in many mines — could be ignited by an open flame. Still, they needed some illumination, both to extract coal and to check on their surroundings in order to spot structural weaknesses in the shafts, so they risked candlelight after an overman checked the workings for gas. First, the overman lit a trimmed and clean candle on the floor and placed his palm in front of it so that he saw only the spire of the flame. Then he raised the candle slowly toward the ceiling of the mine, where firedamp — lighter than air — collected. If it was present, the tip of the flame would turn blue. "This spire increases in size and receives a deeper tinge of blue, as it rises through an increased proportion of inflammable gas, till it reaches the firing point," explained an account of the time. "But the experienced collier knows accurately enough all the varieties of *shew* (as it is called) upon the candle, and it is very rarely fired upon, excepting in cases of sudden discharges of inflammable gas."

In the best circumstances, when the overman found firedamp, he left the mine and then — so as to make it safer for work — ignited the gas by lowering a lighted candle or coal-filled iron basket down the shaft. But if he detected firedamp far inside the workings, he had no choice but to send down a man to ignite it: "Clad from head to foot in rags soaked in water, [the man] would crawl along the underground way holding in front of him a long pole at the end of which was a lighted candle. When the explosion occurred he would fling himself, face downward, on the floor, and so, with good fortune, he might escape the flame which shot along the roof above him." The man was sometimes called a penitent.

In spite of such efforts, miners thought of explosions, and the human injuries and deaths that accompanied them, as inevitable. The history of the mines is also the history of the dead, the burned, and the injured. As one account attests, "Everything in the way of the blast was thrown out at the mouth to the estimated height of 200 yards in the air. Most of the pitmen, having just in time discovered the danger, were drawn up, and escaped unhurt; but some boys, and one man, who were left behind, lost their lives." Another account tells of four men who

were about three hundred yards from the shaft, when the foul air took fire. In a moment it tore the wall from end to end; and burning on till it came to the shaft, it then burst and went off like a large cannon. The men instantly fell on their faces, or they would have been burned to death in a few moments. One of them, who once knew the love of God (Andrew English), began crying aloud for mercy; but in a very short time his breath was stopped. The other three crept on their hands and knees, till two got to the shaft and were drawn up; but one of them died in a few minutes. John M'Combe was drawn up next, burned from head to foot but rejoicing and praising God. They then went down for Andrew; whom they found senseless: the very circumstance which saved his life. For losing his senses, he lay flat on the ground, and the greatest part of the fire went over him.

Miners and mine owners were always looking for alternatives to candles. Although miners' candles were exceedingly small — up to sixty to the pound, for it was believed a small candle might prevent the ignition of firedamp — everything thought of as a substitute for them provided less light than even those slim solitary tapers. It's almost inconceivable now to imagine how slight and shifting was the illumination miners worked by so far below the earth's surface. One device, a flint mill, required boys to accompany the miners down the shafts. Each boy worked a mill, which might be strapped to his leg or hung from his neck. It was made of a steel disk set in a small steel frame and a handle attached to a spur wheel, which turned the disk. The boy held a piece of flint against the disk as he rotated it so as to produce streams of sparks for the miner to work by. The sparks were usually too cool to set off the gas, but not always.

And if miners couldn't use even a mill, they had little else to rely on for illumination. When a flint mill at the Wallsend Colliery caused an explosion that killed nine miners, "work was continued in the shaft without it and with the greatest difficulty. For some time it was performed in total darkness, aided only by light reflected from the surface by means of a mirror during periods of sunshine." Perhaps the strangest form of light was used in the Tyne mines, known to be "gassy" or "fiery." There colliers "sometimes tried to carry on their work by the feeble light of phosphorous and putrescent fish."

The first practical miners' safety lamps were developed around 1815, and the one devised by Sir Humphry Davy, later head of the Royal Society in London, proved to be the most popular. Davy enclosed a flame within a wire mesh cylinder, which distributed the fire's heat and prevented the air beyond the lamp from reaching the ignition temperature of firedamp. Although his lamp was quickly put into wide use, it didn't slow the number of mine deaths. Because of the mesh, the Davy lamp shed only about one-sixth the light of a common taper, so miners often continued to work by candlelight as well. The use of safety lamps also encouraged men to work deeper in the mines and open up more fiery seams. As a result, the mines became even more dangerous. The inventors of safety lamps, one mining historian suggests, "had provided the miner with a weapon of defense: armed with it he was led forward to meet fresh perils. They had sought to bring security of life: they achieved an increase in the output of coal."

By the time Davy developed his safety lamp, an increase in the output of coal had become essential. Not only was the Industrial Revolution speeding up, but coal gas possessed an increasing value. In addition to illuminating the workrooms in factories, gas was illuminating streets and shops and homes in the city of London. Bringing gaslight beyond the factories had required the sustained effort of its promoters, who had to overcome opposition from whale oil and tallow interests and the skepticism of some prominent scientists. Sir Humphry Davy himself thought the idea so absurd that he asked "if it were intended to take the dome of St. Paul's for a gasometer." Five years after Murdoch successfully lit the Soho forge, gas streetlamps made their first modest appearance. In 1807 a section of Pall Mall was outfitted with lamps to celebrate the king's birthday. It would be another five years before German immigrant and entrepreneur Frederick Albert Winsor (born Friedrich Albrecht Winzer) established the world's first gas lighting company, the Chartered Gas Light and Coke Company in London.

Winsor knew of Lebon's thermolampe and envisioned the home system writ large for an entire neighborhood. As Wolfgang Schivelbush notes, "Winsor was not the original inventor of gas lighting. . . . But he established the concept that allowed gas lighting to make the transition from individual to general use: the idea of supplying consumers of gas from a central production site by means of gas mains." Winsor's company, with its single gasometer, delivered gas for street lighting, commercial establishments, and wealthy homeowners in Westminster, Southwark, and the surroundings, including Westminster Bridge. Its brilliance and relative cleanliness was immediately apparent and appealing. Gaslight, it was claimed, shed "a brightness clear as summer's noon, but undazzling and soft as moonlight. . . . Those who have been used only to the brilliancy of oil and candle-light, can have no adequate idea of the effect of an illumination by gas. It so completely penetrates the whole atmosphere, and at the same time is so genial to the eyesight, that it appears as natural and pure as daylight, and it sheds also a warmth as purifying to the air as cheering to the spirits."

Once established, gaslight spread quickly throughout London. By the early 1820s, nearly fifty gasometers and several hundred miles of underground gas mains supplied more than forty thousand public gas lamps for the streets. The lamplighters made their rounds with relative ease, using a lighted oil lamp on the end of a pole. "I foresee in this . . . the breaking up of our profession," a lamplighter in a Charles Dickens story would soon proclaim. "No more polishing of the tin reflectors[;]. . . no more fancy-work, in the way of clipping the cottons at two o'clock in the morning; no more going the rounds to trim by daylight, and dribbling down of the *ile* on the hats and bonnets of the ladies and gentlemen, when one feels in good spirits. Any low fellow can light a gas-lamp, and it's all up!"

In the intimate spaces of home, this strange new light may not have required the same daily attention as did oil lamps and candles, but it had its drawbacks. Its larger flame produced considerable soot and an acid residue that destroyed fabric and wallpaper, and it consumed so much oxygen that people suffered headaches in poorly ventilated rooms, although as time went on, gas chandeliers eventually contained their own ventilation systems. But perhaps more important, with the advent of gaslight people had to reimagine how light would inhabit their homes. Light's abstract future had begun: there was nothing to tend, no wick to see consumed, no melting wax or reservoirs of oil drawing down. The size of the flame could be controlled by a switch and did not waver, flicker, or gutter. It not only stood upright but shot out of the core sideways or upside down, in the shape of a fish tail, a bat wing, or a fan. It was not to be doused with water or extinguished with breath. Fire itself seemed to travel through the pipes. "It was strangely believed that the pipes conveying the gas must be hot!" exclaimed engineer Samuel Clegg. "When the passages to the House of Commons were lighted, the architect insisted upon the pipes being placed four or five inches from the wall, for fear of fire, and the curious would apply the gloved hand to the pipe to ascertain the temperature."

Not only had the nature of the flame itself changed; until gas arrived, light — however meager — had always been one's own and self-contained within each dwelling. Gaslight divided light — and life — from its singular, self-reliant past. All was now interconnected, contingent, and intricate. When people installed gas, they gave up control of light to an outside interest; they no longer purchased candles or oil and carried it home. Rather, their consumption was registered by a meter, they purchased their fuel by the cubic yard, and it was delivered as it was consumed. Their homes were connected to their neighbors' homes, to the homes of strangers, to factories, and to the streets in a shared fate. It marked the beginning of the way we are now, with our nets of voices, signs, and pulses, with power subject to flickers and loss we can't do anything about.

Although for decades gaslight remained the province of better neighborhoods, people throughout the city suffered the streets being dug up for the laying of lines, and along the lines and at the lampposts, gas leaked from ill-fitted joints and seams and from accidental ruptures. Explosions flattened buildings, sent bricks and debris flying, and killed and maimed workers, householders, pedestrians, and shoppers in nearby bakeries and butcher shops. Most affected were the neighborhoods, often among the poorest, that had to endure the presence of gasworks, with their enormous storage tanks looming above the surrounding buildings and their furnaces belching a dense, foul smoke that permeated everything with a sulfurous stench. The gasworks contaminated nearby soils and subsoils with ammonia and sulfur, polluted water supplies, and drove the surrounding area into decline. One critic of the time noted: "Wherever a gas-factory — and there are many such — is situated within the metropolis, there is established a centre whence radiates a whole neighbourhood of squalor, poverty and disease. No improvement can ever reach that infected neighborhood — no new streets, no improved dwelling, not even a garden is possible within a circle of at least a quarter of a mile in diameter, and not so much as a geranium can flourish in a window-sill."

Manufacturers insisted that the smell of gas was good for one's health, but court testimony proved otherwise: "Mr. Arabin, deposed, that he was an upholsterer, residing . . . about two hundred yards from the building belonging to the Gaslight Company. . . . He observed something daily issuing from the establishment exceedingly offensive: it was a kind of smoke producing a saline effluvia, which operated upon his senses, and considerably affected his respiration: the smell was of a sour and acrid nature." According to another witness, "When the effluvia was abroad, he could not open his windows. . . . His own lungs were hurt, and there was a certain nausea produced upon the stomach. A taste was also continually in his mouth, like sulphureous acid. There was an immense quantity of smoke proceeding from excessive large fires, and when these appeared to be at work, he was compelled to close up his doors and windows." The testimony of a third witness echoed the sentiments of the other two: "*Thomas Edgely* is a coal-merchant, and has a wharf adjoining the gas light manufactory, and from which there is a constant stench. . . . Never remembers to smell anything so offensive in his life; even the *coal-heavers* complain and are sickened by it. Believes it is no easy matter '*to turn a coal heaver's stomach.*'"

Fear of gasometer explosions became part of the anxiety of the age. The *Times* of London suggested that "at present it is clear every gasometer is a powder-magazine, and to have a gas manufactory near Westminster Abbey, St. Paul's, or one of the bridges, is much the same as if we were to store our gunpowder on the Thames Embankment." The fear wasn't allayed over time, for the gasometers would only become more prominent and increase in size as the century progressed. Within the works, men — dwarfed by the furnaces and in the flare of fire — shoveled coal into the retorts. We can see them there in a moment of respite in Gustave Doré's 1872 wood engraving *Lambeth Gas Works:* clustered, exhausted, dressed in rags. Behind them, the even courses of the brick walls, the arch with its keystone — twice the height of the men — and the gas pipes remain unassailably solid, as do the men in the far distance working the furnaces: stiff-backed, stiff-armed, disciplined, stoking the fires in mechanical unison, seeming to be part of the machine. The men at rest aren't sheltered by the immensity — they are dominated by it, and in the uncoordinated moment of their exhaustion, with their shoulders hunched, their ragged clothes draped over them, they find no relief in not being part of the machine: they have been defeated by it.

For all its complexities, gaslight proved to be a remarkable success in London, and that success led to its rapid establishment in other British cities and towns. Historian Stephen Goldfarb notes: "In 1821 no town in the United Kingdom with a population of more than 50,000 was without a gas company; by 1826 only a few towns over 10,000 lacked gas companies; and by mid-century a 'vast majority of towns with a population greater than 2,500 possessed gas companies.'" Across continental Europe and in America, where economies were still based on wood, gaslight appeared later, progressed more slowly, and remained largely an urban system. "Paris was illuminated in 1814 by 5,000 [oil] street lamps, serviced by 142 lamplighters. . . . In 1826 there were 9,000 gas burners in Paris; in 1828 there were 10,000." In the United States, Baltimore was the first city to adopt limited gas lighting, in 1817. Philadelphia and New York experimented with gaslight at the same time, but it failed to take hold, in part because of opposition from tallow manufacturers. New York's first gaslights appeared in 1825, Philadelphia's in the 1830s.

Robert Louis Stevenson wrote in praise of gaslight:

The work of Prometheus had advanced another stride. Mankind and its supper parties were no longer at the mercy of a few miles of sea-fog; sundown no longer emptied the promenade; and the day was lengthened out to every man's fancy. The city-folk had stars of their own; biddable domesticated stars. . . . It is true that these were not so steady, nor yet so clear, as their originals; nor indeed was their lustre so elegant as that of the best wax candles. But then the gas stars, being near at hand, were more practically efficacious than Jupiter himself. It is true, again, that they did not unfold their rays with the appropriate spontaneity of the planets, coming out along the firmament one after another, as the need arises. But the lamplighters took to their heels every evening, and ran with a good heart.

Under gaslight, the true stars started to fade away. "Paris will be very beautiful in autumn," wrote Vincent van Gogh from Arles in 1888 in a letter to his brother, Theo. "The town here is *nothing*, at night everything is *black*. I think that plenty of gas, which is after all yellow and orange, heightens the blue, because at night the sky here looks to me — and it's very odd — *blacker* than Paris. And if I ever see Paris again, I shall try to paint some of the effects of gaslight on the boulevard." Van Gogh seems to be referring to what might be sky glow, that aspect of light pollution in which the night sky appears purplish in the glare of multitudinous lights.

By mid-century, the long view of a gaslit city could appear simply enchanting. "The whole of Paris is studded with golden dots," a guidebook to the city observed, "as closely as a velvet gown with golden glitter. Soon they wink and twinkle everywhere, and you cannot imagine anything more beautiful, and yet the most beautiful is still to come. Out of the dots emerge lines, and from the lines figures, spark lining up with spark, and as far as the eye can see are endless avenues of light." A closer look, however, often told other stories, for the allure of an illuminated city at night is much more than a matter of streetlights alone, which are simply the strict lines of civil order and, by themselves, markedly utilitarian whether gas or oil. A city night thrives in myriad lights — shop windows, signs, theater entrances, taverns, homes — and in the gaslit neighborhoods, the brightness of all of the illuminated places increased exponentially, which in turn fed the vitality of the streets. People who lived in gaslit neighborhoods grew accustomed to the brightness and often felt safer in their larger illumination. Those districts still dependent on feeble, messy oil lamps — most often working-class and poor neighborhoods — were another country now, a place into which the well-to-do might be more reluctant to venture, as if the gloominess of oil lamps marked the edges of their territory.

By the gaslight era, too, a rising middle class had more leisure time in the evening and more money to spend. The evening became the consumer's hour, with the advent of window-shopping as a pastime. These were the hours of glass, which had been clarifying ever since the sixteenth century, when small panes of it first began to replace muslin and oilpaper in windows. Now the glass in shop windows — no longer composed of small panes — was one large plate, which gaslight, unlike oil lamps and candles, suffused with light, illuminating the still lifes within. Steady and mute, it fell upon sequined dresses, wool coats, and silk ties; on watches and necklaces perched on folds of velvet; on fabrics, perfumes, soaps, silver candlesticks, Chinese porcelain, Indian spices, cheeses, and meats.

Whereas plate glass gave a view into shop interiors, in noisy cafés light seemed to ramify endlessly off glass chandeliers, bottles of whiskey and absinthe, stemware and tumblers. Mirrors magnified that light even further. "During the day, often sober; in the evening, more buoyant when the gas flames glow," wrote Karl Gutzkow of Paris. "The art of the dazzling illusion is here developed to perfection. The most common-place tavern is dedicated to deceiving the eye. Through mirrors extending along walls, and reflecting rows of merchandise right and left, these establishments all obtain an artificial expansion, a fantastical magnitude, by lamplight."

The theater, too, was transformed by gaslight — and lime-light, which was used first as a signal by surveyors and then adapted for the theater in the 1830s. Not only had the candle-snuffer with his interruptions become a thing of the past, but light could now be dimmed and heightened with ease, which allowed for more sophisticated lighting effects, and the stage could be more intensely lit than the rest of the theater, which formally isolated the performance from the audience. Actors had to adjust to the new light. "The new mode of illumination made it rather difficult for old-line declamatory actors . . . to practice the tricks of their trade," notes theater historian Frederick Penzel. "All of a sudden, gestures seemed overbroad, and facial expressions seemed greatly exaggerated. What had apparently worked before murky candlelights was no longer effective before the gaslights. Even the makeup appeared garish. Things only half-seen before were now totally revealed, and all had to be toned down."

Gaslight also transformed the crowds walking the streets: darting eyes, staring eyes, averted hooded eyes; myriad sounds and colors; confinement and freedom — all became illuminated. What was a walker but "a *kaleidoscope* equipped with consciousness," according the street a soul, according it the power to take one's own away? Humanity at night had become the sea. "As the darkness came on, the throng momently increased," Edgar Allan Poe wrote; "and, by the time the lamps were well litten, two dense and continuous tides of population were rushing past the door." But the streetlights, as Poe saw it, also illuminated different aspects of human nature:

As the night deepened . . . not only did the general character of the crowd materially alter (its gentler features retiring in the gradual withdrawal of the more orderly portion of the people, and its harsher ones coming out into bolder relief, as the late hour brought forth every species of infamy from its den,) but the rays of the gas-lamps, feeble at first in their struggle with the dying day, had now at length gained ascendancy and threw over every thing a fitful and garish lustre. All was dark yet splendid.

Buoyant, frivolous, expansive, uncontainable humanity: light seemed not only to extend the hours of the day but also to have created life out of absence and to have allowed for different qualities in human nature to have their say. Surely, the medieval cities lay buried under paving stones, and the ancient perimeter gates had been lost in the sprawling reaches. "Night" — that one old, taut syllable once uttered with fear and apprehension — no longer sufficed. In the middle of the nineteenth century, a new word was minted: "nightlife."

But what happened when the old night returned, as it inevitably did when a gas explosion or a gasworkers' strike occurred? Such instances didn't darken an entire city — most municipalities of any size were served by various competing gas companies, each having contracts for certain districts. Also, when a retort shut down or an explosion occurred, a vestige of gas remained in the system, so the lights continued to shine for a while, then dimmed before disappearing completely. Still, even a few hours in darkness caused major alarm — and more alarm as the century progressed and the dependence on gaslight grew. A *New York Times* account titled "Bereft of Light" detailed the events in a neighborhood in the wake of a gas explosion at the Metropolitan Gas Works on December 23, 1871, which darkened the area between Thirty-fourth and

Seventy-ninth streets. The explosion shattered windows, sent bricks flying, stopped clocks, and startled horses. There was a fire, put out in several hours, which injured a fireman. But by far the most newsworthy part of it all was the anxiety:

> Some rushed about from house to house while the more thoughtful ones besieged the Police Stations. . . . The store-keepers lighted up their shops with candles and lamps as well as they could, but yet there was no real show of light. . . . They bought pounds of candles and made temporary rustic candle-sticks of fruits and vegetables for the purpose of showing off their stock, but it was of no avail; the citizens seemed to be too much concerned about the loss of gas to think of spending a cent on fruit or anything else . . . and it is probable that for many years so many of the citizens of the City have not retired to rest so early in the evening. The various bank officers were in a great state of agitation. They rushed to the Police Precincts on the first alarm, and, having obtained strong guards, immedi-ately set to work to arrange kerosene-lamps with reflectors over them around the safes.

For those hours, the well-heeled were more exposed and helpless than those living by oil and candles. No longer privi-leged in the night, anxious for some power beyond them to restore the lamps so that life could fold back into its hurry, they could only wait in the midst of the old quiet, where light had circumference again.

5

TOWARD A MORE
PERFECT FLAME

THE FIRST DECADES OF the nineteenth century brought marked changes not only for those living within the sphere of gaslight but also for households that continued to rely on oil lamps and candles alone. Manufactured candles became cheaper and improved so much in quality that even the smallest flame in an ordinary home possessed some of the properties of beeswax and spermaceti. Part of the improvement could be attributed to plaited wicks and wicks impregnated with boric acid, which helped to diminish guttering, but much of it had to do with the substance of the candles themselves. Commercial tallow manufacturers developed a way to refine animal fat so that it no longer smoked or stank as it burned. The famed scientist Michael Faraday explained the process in *The Chemical History of a Candle*:

> A candle, you know, is not now a greasy thing like an ordinary tallow candle, but a clean thing. The fat or tallow is first boiled with quick-lime, and made into a soap, and then the soap is decomposed by sulphuric acid, which takes away the lime, and leaves the fat rearranged as stearic acid, while a quantity of gly-

cerin is produced at the same time. . . . The oil is then pressed out of it; . . . how beautifully the impurities are carried out by the oily part . . . and at last you have left that substance, which is melted, and cast into candles.

By mid-century, candles were also manufactured from paraffin, which was derived from the distillation of bituminous shale. An account of the time describes paraffin as "brilliantly white, inodorous, and tasteless. It resembles spermaceti in its silky feeling and physical structure . . . [and] derives its name from two Latin words, *parum*, little or none, and *affinis*, affinity, because of its complete neutrality and great stability. . . . It gives a powerful, clear flame, without soot." The name seems to have said it all — here was light unencumbered by noxious smells and smoke, light in little need of tending, that could simply shine constant, clear, and bright. These developments seemed to signal the end of earthbound light, for at last the common candle had distanced itself from the barnyard and slaughterhouse, from blood and sinew and bone, and there would be no going back. Herman Melville, writing in the mid-nineteenth century, proclaimed that even in the bowels of a whaling ship, it seemed an outlandish thing that "mortal man should feed upon the creature that feeds his lamp."

Common lamp fuel, too, underwent changes. In the 1830s, "burning fluid," a mixture of camphene (distilled from turpentine) and alcohol, often called simply "camphene," arrived on the American market. Although it was thin and light, and so traveled quickly up a wick, it had a low flash point — the temperature at which oil gives off enough vapors to spontaneously ignite — which made it volatile. Because any spark or excessive heat could lead to an explosion, the flame of a camphene lamp needed to be kept at a distance from its fuel reservoir. Whale oil and grease lamps were designed with metal wick tubes that extended down into the font so that some heat could warm the oil and make capillary action more efficient. By contrast, lamps that burned camphene had long, narrow wick tubes that extended upward from the font and angled away from the center of the lamp so as to keep the flame at a distance from the fuel reservoir. You also couldn't blow out a camphene flame — a stray spark might ignite the reservoir — so burners often had extinguisher caps attached to them. Even with all these precautions built into lamps, there were thousands of deaths and injuries from explosions each year. According to the *New York Times* in 1854, for

> any common use about the house, it must be confessed after many years of costly experience, that "burning fluid" is not safe.... Nobody will stop using it on this account, however.... Better, if you are coming to your senses, reader, take your lamps at once to the lamp-store and have them all refitted for oil, or something else than "camphene" in any of its forms. If not, be consistent, and stuff your pin-cushions and mattresses with gun-powder, and buy a rattlesnake as a pet for your growing boy to play with.

Why burning fluid was so popular is something of a mystery. Though less expensive than sperm oil, it still retailed for about 25 cents per quart in the early 1850s. And while those of the day said that it produced a brilliant white flame — enough, perhaps, to risk the dangers — historian Jane Nylander claims that "a burning fluid lamp produced a dimmer light than a tallow candle or a single wick whale oil lamp." It may have been that burning fluid was simply the new thing and at a clear remove from age-old animal fuel.

Lamps and candles, whatever their fuel, were now easier to light because people no longer had to borrow fire from an existing flame or coals, or resort to a tinderbox. Even in the early nineteenth century, the few alternatives to such methods were for the wealthy alone, who might carry phosphoric tapers, or "Ethereal Matches," with them. These were short strips of paper tipped with a bit of phosphorus, and each was contained within a thin glass vial. When the user broke the glass, the phosphorus burst into flame. Such matches came with their share of injuries, since they might also burst into flames if the vial was accidentally broken.

In 1826 Englishman John Walker developed what would eventually evolve into the common match. He made his "friction-light" by dipping wooden splints into a paste of potassium chlorate, starch, antimony sulfide, gum arabic, and water. He dried the splints and then ignited them by nipping them between folded sandpaper. Early matches sparked and stank, and Lucifers — as the matches came to be called — carried a warning: "If possible avoid inhaling gas. Persons whose lungs are delicate should by no means use Lucifers." One Parisian exclaimed, "The chemical match is, without doubt, one of the vilest devices that civilization has yet produced. . . . It is thanks to this that each of us carries around fire in his pocket. . . . I . . . detest the permanent plague, always primed to trigger an explosion, always ready to roast humanity individually over a low flame."

Eventually, matches were coated with white phosphorus, and they became safer for those who carried them, but not for those who made them. Match makers who were exposed to phosphorus vapor for long periods of time commonly suffered from painful, disfiguring phossy jaw, which was fatal. Deposits of phosphorus in the jawbone would eventually begin to abscess, and the bone would rot away. The sufferer would then die from organ failure. Although less toxic red phosphorus began to replace white phosphorus in the mid-nineteenth century, white phosphorus was still used in the production of "strike-anywhere" matches until early in the twentieth century.

The light of even common whale oil lamps improved in the early nineteenth century, as a wide array of new lamp designs came on the market in the wake of Ami Argand's revolutionary invention of the tubular wick in 1784. Wall lamps, table lamps, night lamps, student lamps, and chandeliers were crafted of pressed glass, pewter, silver, iron, brass, nickel plate, and japanned tin. The more complex models attempted to improve on the delivery of thick whale or colza (rapeseed) oil to the wick so as to eliminate the obscuring reservoir of the Argand lamp. The Carcel lamp used a clockwork pump to feed fuel; the moderator lamp had a strong spring that pressed down a piston, which squirted oil up a narrow tube; and the astral lamp featured a ring-shaped oil font. The complexity of such lamps, and their prodigious fuel requirements, meant they were out of economic reach for people of limited means, but eventually even manufacturers of the most simple single-burner lamps adopted hollow wicks and glass chimneys, which increased oxygen flow and stabilized flames. The new lamps often had double or even triple wicks, which meant that one lamp could burn with different intensities, a preliminary version of the contemporary three-way bulb.

But the most significant improvement to the lamp since Argand's invention came in the second half of the nineteenth century, with the advent of kerosene. "We dreamed of the lamp which gives luminous life to dark matter," wrote Gaston Bachelard of the kerosene lamp. "How could a dreamer of words not be moved when etymology teaches him that petroleum is petrified oil? The lamp makes light ascend from the depths of the earth." "Rock oil" had been gathered from seeps for thousands of years and was used in its crude form, mostly as a lubricant or medicine, all over the world. North American Indians collected surface oil by soaking it up with blankets. They then applied the oil as a salve or used it to waterproof their canoes.

In 1849 Canadian geologist Abraham Gesner developed a way to extract what he called "kerosene" from asphaltum — a type of mineral pitch — and subsequently oil refiners discovered that they could use Gesner's process to produce kerosene from petroleum. Its production wouldn't become commercially viable until 1859, when Edwin Drake drove the first successful oil well in Titusville, Pennsylvania, which yielded a reliable supply of petroleum for refining.

Housewives adopted kerosene for myriad uses: they wiped it over bedding and on kitchen walls and screen doors to keep bugs away; poured it on anthills and cleaned flyspecked brass with it; used it to clean their porcelain sinks, marble washbowls, windows, and cookstoves; removed rust, fresh paint, and grease from their graniteware with it; and added it to hot starch to keep the starch from sticking to clothes. But they valued kerosene most highly for light. Although the quality of a kerosene lamp's flame varied with the quality of the fuel and the size and cleanliness of the lamp and the wick, at its best it burned clear, hardly smoked, and was relatively odorless. One kerosene lamp burned as brightly as five to fourteen candles.

Unlike animal fuels, kerosene would not spoil on the shelf over time, and quality kerosene was considered safe and stable because it had a high flash point. It also was light — much lighter than whale oil and colza — so it required no clockwork or pistons to travel up a wick. Since there was not yet competition from the internal combustion engine for petroleum supplies, it was economical — cheaper than either whale oil or gas. By 1885 it was claimed that this new fuel "could supply a family's needs for about ten dollars a year 'while it was not uncommon for the gas bill of the more well-to-do householders to run that much per month.'" Kerosene was, as William O'Dea notes, "the kind of oil people had dreamed about for centuries."

The immediate demand for kerosene ushered in the age of oil. In the months following Drake's first well, land prices around Titusville shot up, and the population multiplied many times. Within a year, numerous refineries in the oil regions of Pennsylvania and in Pittsburgh began operation. Early shipments went to New England and the Mid-Atlantic States. After the Civil War, kerosene spread into the Midwest and more slowly penetrated the postwar South. Eventually, more than half of the American supply was shipped overseas to Europe and Russia, which established the fortune of John D. Rockefeller and Standard Oil. Still, the supply could appear fragile. During those first decades of drilling, all the kerosene produced was derived from the Pennsylvania oil fields, and the extent of the reserves was unseen and unknown. Yet oil had already become so essential to modern life that in 1873 the *Titusville Morning Herald* proclaimed: "The production of petroleum has now become of such commercial and social importance to the world that if it were suddenly to cease no other known substance could supply its place, and such an event could not be looked upon in any other light than of a widespread calamity."

There was little place left for whale oil in such a world. Within a year of Drake's oil rig, kerosene had replaced it as the popular fuel. In truth, though, the northeastern American whaling fleet was already in decline by 1859. Although sperm whales had not become extinct, they had grown scarce by the latter part of the nineteenth century. Consequently, the hunt for them was more time-consuming, arduous, and costly. When the Civil War broke out in 1861, cautious northern merchants kept their ships tied up in port — it wasn't worth risking capture by Confederate cruisers — and after several years, the hulls of the whaling ships began to rot in the wharves. Once the war was under way, in an effort to blockade Charleston and Savannah harbors, the Union purchased forty old whalers, loaded them with stones, and sank them. When peace arrived in 1865, the fleet was a fraction of what it had been, and since a good share of their market had been lost to kerosene, most merchants chose not to replace their old vessels. The ships that did set out were obliged to take on more and more risk in order to fill their holds, sailing longer into the northern winter and escaping the freeze-up by increasingly narrow margins.

It was also true that the northeastern fleet, in an age of steam, had failed to modernize, which was essential, since the whaling grounds had shifted largely to the Arctic, where conditions for both seamen and ships were brutal. The rudders of sailing ships had to be kept free of ice, and in frigid conditions ice formed on the riggings, and the ships were in danger of capsizing from the weight. When, in the early winter of 1871, thirty-two ships became locked in Arctic ice — the crews surviving by making their way to vessels in open water — the loss of more ships meant that the hunt, for the northeastern fleet, was nearly finished.

Although animal fuel would not again feed more than an occasional lamp, demand for baleen and other products made from whale blubber — margarine, soap, lubricants — continued. Faster steam whalers that shipped out of San Francisco and northern Europe hunted species that could not have been captured under sail alone. Although the sperm whale had grown scarce, it was still hunted as well, for its oil retained its lubricating qualities in extreme temperatures. Sperm oil would grease the machines of the industrial age long after the last whale oil lamp went out. Prior to the 1982 international moratorium on whaling, sperm oil — harvested from a mammal that could dive to depths of more than four thousand feet, the deepest of all mammals — would lubricate the precision instruments on spacecraft.

For all its popularity, kerosene had some drawbacks. In the first decades of its manufacture, there was little regulation of the supply, and unscrupulous traders adulterated it with benzene or naphtha, which lowered the flash point and made it more volatile. The members of the Buffalo, New York, Board of Trade noted:

> The country has been flooded with all sorts of compounds and mixtures and greases that pretend to do everything and accomplish nothing. In refined oils particularly, low-test oils and fluids have been thrown on the market by unprincipled men, with perfect impunity, throughout the country. . . . The inspector may do his duty, and brand the oil — the legal test; but how easy it is for the refiner or the dealer to add a few gallons of death-dealing naptha [*sic*], for *profit's sake* and thereby endanger lives and property.

Responsible housewives had to be vigilant about the quality of the oil they purchased. Catharine Beecher and Harriet Beecher Stowe, in their 1869 guide to domestic science, *The American Woman's Home*, advised:

> Good oil poured in a teacup or on the floor does not easily take fire when a light is brought in contact with it. Poor oil will instantly ignite under the same circumstances, and hence,

the breaking of a lamp filled with poor oil is always attended by great peril of a conflagration. Not only the safety but also the light-giving qualities of kerosene are greatly enhanced by the removal of these volatile and dangerous oils. Hence, while good kerosene should be clear in color and free from all matters which gum up the wick, and thus interfere with free circulation and combustion, it should also be perfectly safe.

Even with quality kerosene, the simplest lamp found in the most modest of homes required meticulous daily attention. Only a well-cleaned lamp would give off good light, and a poorly trimmed wick made for a flickering and smoky flame, which left soot on the chimney and sent soot throughout the house. Indeed, the ritual of spring-cleaning was largely a response to a winter's worth of soot from hearths and lamps. But daily cleaning was also a matter of safety. In the late nineteenth century, in the United States alone, five thousand to six thousand people a year died in lamp accidents. Although many of these were due to adulterated oil, clumsiness, and carelessness — spills and breakages, or someone leaving a lamp too near to curtains or bedclothes, failing to lower the wick before blowing out the light, or trying to extinguish a lamp by blowing down the chimney — inattentive or inexperienced housekeeping increased the danger. If the burner was dirty, the lamp might overheat the chimney and break the glass. If the oil in the reservoir was too low, the vapors could ignite when someone carelessly jarred the lamp. A Connecticut newspaper, the *Willimantic Chronicle*, often reported lamp accidents due to "exploding lamps":

Wed., September 1, 1880: The house of George Leavens in Danielsonville came near being consumed by fire by the explosion of a kerosene lamp last week. . . . Wed., August 29, 1883: Early Saturday morning the body of Simon B. Squires was found in the back yard of the Southport National bank, Southport, burned in a shocking manner. It is thought he rose during the night when his lamp exploded and set his clothes on fire. . . . Wed., April 23, 1884: Mrs. Mary McGoldrick, aged 73 years, of New Haven, Ct., and Emma O'Brien, aged 3 years, of Erie, Pa., were yesterday burned to death by the explosion of kerosene lamps.

The authors of *The Woman's Book*, a guide to household management published in 1894, went into a lengthy, precise discussion about the cleaning of lamps:

> There is as much wit goes to the care of lamps as to the boiling of eggs. In the first place they should receive due attention every day. . . . Carry the lamps to the kitchen or pantry and set them down upon double-folded newspapers. If they have porcelain shades, wipe these. . . . Should they need washing, put them into a basin of hot water which you have softened with a little ammonia or borax. . . . This done turn up the wicks of the lamps and with a bit of stick or match scrape off the charred edges. . . . Remove the rims that surround the burners and wipe them off with old flannel. . . . Now fill the lamps and do it carefully. . . . Wipe the outside of the reservoirs after you have filled and closed them, that the persistently percolating oil may have not unnecessary encouragement to exude. Be very sure that no drops of oil have trickled down upon the outside of the lamps. . . . Give a final rub to the outside of each lamp, replace chimney, rim, and shade, and thank Fate that this, one of the least pleasant of the housekeeper's duties, is done for the day.

However much work the lamps proved to be, for those living in towns, villages, and farms beyond the reach of gaslight, kerosene brought significantly more light into homes. Not only was each lamp brighter, but the low cost of fuel also encouraged people to use lamps more frequently and to purchase more lamps. Kerosene goods, such as reservoirs, wicks, and chimneys, became an advertised standard in catalogs and general stores. With the kerosene lamp's ease came a certain thoughtlessness, and perhaps an appreciation for the beauty of the flame itself. Certainly, people could read or knit with far less strain, and work more steadily by its light. Since by the second half of the nineteenth century, enclosed wood and coal stoves had begun to replace open hearths, the kerosene lamp — the last open fire in the home — often became a gathering place for the family in the evening.

Even some city people connected to gas lines reserved gas for the utilitarian spaces of their homes, such as hallways and kitchens; they continued to use oil lamps in more intimate drawing rooms and bedrooms. Historian Wolfgang Schivelbush argues that it wasn't gaslight's palpable drawbacks — the soot and the bad air — that made most people hesitant to fully give in to gas. Rather, it had to do with the industrial source of its flame and all that implied of a connection with, and a dependence on, the brick and gray life looming beyond, the cinders and ash settling over cities and towns. And more: "By keeping their independent lights, people symbolically distanced themselves from a centralised supply," Schivelbush notes. "The traditional oil-lamp or candle in a living-room expressed both a reluctance to be connected to the gas mains and the need for a light that fed on some visible fuel."

Some simply preferred the modest flame of the old light: "I boldly declare myself the friend of Argand lamps," stated one Parisian in comparing them to gas lamps; "these to tell the truth are content with shedding light and do not dazzle the eyes." Perhaps, for city dwellers, as the oil lamp began to take its place as part of the past — its notes diminishing as others sounded and strengthened — its intimacies seemed all the more desirable, and people instinctively clung to its lingering form, the ghost in the mist. "It seems there are dark corners in us that tolerate only a flickering light," wrote Gaston Bachelard. That flickering was a link to the light at the beginning of human time: the kerosene lamp was the apotheosis of the tallow cupped in limestone at Lascaux, the last self-tended flame.

PART II

You turn the thumbscrew and the light is there.

—*New York Times*, September 5, 1882

6

LIFE ELECTRIC

HUMAN LIGHT HAS ITS SOUNDS — of a match struck and a candle flame muttering in a draft, of a stopcock turning and a gas jet hissing to life or hoarsely damping itself out. Now: the crackle and snap of electricity — for thousands of years a mystery and arriving as light only after ages of isolated experiments, speculation, observations, and discoveries. Light that required a new vocabulary — amps, volts, watts, joules, the galvanic cell. Light without fire, incandescently silent, its switch a "little click [that] says *yes* and *no* with the same voice." It was the harnessing of what has been marvelous at least since the ancient Greeks saw the way amber, when rubbed with a piece of wool, created sparks, so they could only conclude that it, too, had a soul, for "it seemed to live, and to exercise an attraction upon other things distant from it." Amber, which the Greeks believed were the tears of the Heliades, Phaëthon's sisters, who wept so long beside the river where he'd drowned that the gods in their pity turned them into poplars.

The philosopher Thales, who lived around 600 B.C., was the first to mention the sparking of amber in his writings, though its electrostatic qualities were likely already well-known. The

Greeks, it was said, treated gout by standing on electric eels, but whether they used amber for any practical or religious ends is only conjecture, as is the use of ancient batteries, dating to around 200 B.C., found in the vicinity of Baghdad. The five-inch-high clay vessels each contain an iron rod encased in a copper cylinder. One, if filled with vinegar, grape juice, or lemon juice, could have delivered a few volts of power. Archaeologists found needlelike objects near some of the batteries, so perhaps the current was used in acupuncture. Or the batteries may have been connected in series to produce a greater charge for electroplating. Or perhaps statues of idols were wired to them so that small shocks might inspire awe in supplicants.

Electricity's modern path can be traced back to 1600 in London, where Dr. William Gilbert, surgeon to Queen Elizabeth I, noted in his *De magnete* that sparks flew not only from amber but also from glass and precious stones, resin, sulfur, sealing wax, and more than a dozen other substances. He called these substances "electrics," from the Latin word *electrum*, in turn derived from the Greek word for amber, *elektron*. Gilbert died only a few years after the publication of his work, though in succeeding years other scientists, knowing of his findings, extended the list of electrics — among them diamonds, white wax, and gypsum — which remained just a list until Otto von Guericke, mayor of Magdeburg (now in Germany) created an electrostatic machine: a small, solid sulfur globe about six inches in diameter, set in a wooden frame, which he turned with an attached handle. When he both rotated and quickly rubbed his machine, it not only glowed and sent sparks flying; it also attracted light objects.

Guericke noted that electricity could repel things as well as attract them, and to the amusement of friends and visitors, he used his whirling globe to drive feathers across his drawing room, guiding them along until they rested on his guests' noses. For decades afterward, electricity — understood as a "virtue" — would remain largely an enigma that thrived as entertainment. An increased understanding of its properties only inched forward as a result of occasional observation of phenomena between amusements.

In the early eighteenth century, Englishman Stephen Gray established the conductive properties of electricity, having found, after rubbing the bottom of a glass tube, that its cork stopper had become charged. Through his experiments, Gray also discovered the insulating properties of some substances:

> He suspended a long hempen line horizontally by loops of pack-thread, but failed to transmit through it the electric power. He then suspended it by loops of silk, and succeeded in sending the "attractive virtue" through seven hundred and sixty-five feet of thread. He at first thought that the silk was effectual because it was thin; but on replacing a broken silk loop by a still thinner wire, he obtained no action. Finally he came to the conclusion that his loops were effectual, not because they were thin, but because they were *silk*.

With this knowledge, Gray developed his "dangling boy" experiment, which in succeeding years became popular in drawing rooms across England. He suspended a young boy — swathed in nonconducting clothes except for his head, hands, and a few toes — by thick silk ropes. The boy held a wand with a dangling ivory ball in one hand and stretched out his other hand freely. When Gray set an electrified glass tube against the child's bare toes, the boy's hair stood on end, and brass leaf that had been piled on the floor beneath him rose toward the ivory ball, his extended hand, and his face. Gray might then invite members of the audience to stand on some conductive material and touch the boy, whereupon they would receive shocks.

The sulfur globes, and the glass ones that succeeded them, could only produce electricity; the first record of its successful storage dates from 1745. In Camin, Germany, Ewald von Kleist wrote of an experiment in a letter to a friend:

> When a nail or piece of brass wire is put into a small apothecaries' phial and electrified, remarkable effects follow; but the phial must be very dry and warm. I commonly rub it over beforehand with a finger on which I put some powdered chalk. If a little mercury or a few drops of spirits of wine be put into it, the experiment succeeds the better. As soon as the phial and nail are removed from the electrifying glass, or the prime conductor to which it hath been exposed is taken away, it throws out a pencil of flame so long that with this burning machine in my hand I have taken about sixty steps. . . . I can take it into another room, and then fire spirits of wine with it. If while it is electrifying I put my finger or a piece of gold which I hold in my hand to the nail, I receive a shock which stuns my arms and shoulders.

Scientists in Leiden (or Leyden), Holland, refined von Kleist's machine, and thereafter it was known as a Leyden jar. The most elaborate of the jars consisted of a water-filled glass container with an outer and inner coating of metal foil and metal filings at the bottom of the jar. It was capped with a cork or a wooden lid, from which a conductor — a metal rod, usually brass, topped with a metal ball — protruded. A metal chain hung into the jar from the lid. Experimenters could transfer the electric charge from a whirling globe to the protruding ball; the charge traveled down the rod and chain to the water and foil. A Leyden jar could retain its charge for several days, which, as historian Philip Dray notes, allowed experimenters "to move electricity about as part of a graduated process, not merely to see it as the sudden flash that occurred between objects in a friction experiment."

One of the first experimenters in Leiden found that the jar contained enough power to make his whole body quiver. "I advise you never to try [it] yourself," he wrote to a colleague, "nor would I, who have experienced it and survived by the Grace of God, do it again for all the kingdom of France." But many others across Europe and in America did try it in the succeeding decades. Men administered shocks to small animals and birds, to themselves and their wives; they suffered nosebleeds and fevers, convulsions and weakness. Still they experimented. Abbé Jean-Antoine Nollet, at the court of Louis XV at Versailles, in an effort to see how far a shock could travel, sent a charge through 180 soldiers who'd joined hands. He was satisfied to see that they all jumped in unison, and then he tried the experiment on 750 Carthusian monks, who, holding wires between them, formed a line 5,400 feet long. As the abbé sent the current through, they, too, all jumped at the same moment.

Experimenters made bells ring, set rum on fire, and sent sparks shooting around gilded picture frames. They generated "electric kisses" by suspending a young woman in the same way Gray had suspended his "dangling boy." They then invited men from the audience to kiss her on the cheek, and sometimes the charge was significant enough to crack teeth. Still, electricity remained "a vast country, of which we know only some bordering provinces," and its experimenters were thought to be dabbling in a toy science, for no one had yet found a practical application for its power.

Benjamin Franklin, one of the eighteenth century's most tireless "electricians" — a phrase he coined and by which electri-

cal experimenters were then known — was "chagrined a little that we have been hitherto able to produce nothing in this
way of use to mankind." He knew electricity's true power only
too well, having received at least one considerable jolt. "I have
lately made an experiment in electricity that I desire never to
repeat," he explained in a letter to a friend in Boston.

> Two nights ago, being about to kill a turkey by the shock from
> two large glass jars, containing as much electrical fire as forty
> common phials, I inadvertently took the whole through my
> own arms and body. . . . The company present . . . say that the
> flash was very great, and the crack as loud as a pistol; yet, my
> senses being instantly gone, I neither saw the one nor heard the
> other; nor did I feel the stroke on my hand, though afterwards
> found it raised a round swelling where the fire entered, as big
> as half a pistol-bullet, by which you may judge the quickness
> of the electrical fire, which by the instance seems to be greater
> than that of sound, light, or animal sensation.

Franklin advanced the understanding of electricity with
countless experiments and considerable writings on the subject, and he clarified some of its mystery. Philip Dray notes
that Franklin "was the first to discover that the [Leyden] jar's
stored charge was not in the water, as others had believed,
but in the glass. The glass was a dielectric, meaning it stored
and allowed the passage of electricity but did not conduct
it." Perhaps most significantly, Franklin — like Stephen Gray
and Abbé Nollet — suspected that lightning and the electrical
charges they'd created in their experiments were one and the
same substance. The common belief at the time, however, held
that lightning — heavenly fire — was its own distinct phenomenon and a manifestation of the will of God, a belief that may
have been reinforced by the fact that churches and monaster-

ies, with their high steeples and bell towers, were often struck during storms. "There was scarce a great abbey in England which was not burnt down with lightning from heaven," notes a church history of Britain. Many thought such destruction could be warded off by the sounding of church bells during electrical storms, though the practice only served to hasten the deaths of countless bell ringers.

Franklin suggested a new way to ward off such destruction. "There is something . . . in the experiments of points, sending off or drawing on the electrical fire," he wrote. "For the doctrine of points is very curious, and the effects of them truly wonderful. . . . I am of the opinion that houses, ships, and even towers and churches may be effectually secured from the strokes of lightning by their means." When he began to promote the use of lightning rods on buildings, he encountered considerable resistance from church leaders, who claimed the rods were blasphemous and warned that drawing lightning from the sky would cause earthquakes. He was undeterred, however, and his observations of the workings of lightning rods led to his most renowned experiment, which proved that the charges in the heavens and those in Leyden jars were one and the same.

In July 1750, Franklin proposed that a sentry box, large enough to house a man and with a pointed rod rising from it, be built. It would contain an electrical stand which, if it

be kept clean and dry, a man standing on it when such clouds are passing low might be electrified and afford sparks, the rod drawing fire to him from a cloud. If any danger to the man should be apprehended (though I think there would be none), let him stand on the floor of his box, and now and then bring near to the rod the loop of a wire that has one end fastened to the leads, he holding it by a wax handle; so the sparks, if the rod is electrified, will strike from the rod to the wire and not affect him.

In May 1752, before he could conduct his experiment, a French physicist successfully followed his suggestion. The following month, Franklin, knowing nothing of the events in France, carried out a similar experiment with a silk kite, a hemp rope, and a key, which he later detailed:

> As soon as any of the thunder-clouds come over the kite, the pointed wire will draw the electric fire from them, and the kite, with all the twine, will be electrified, and the loose filaments of the twine will stand out every way, and be attracted by an approaching finger. And when the rain has wetted the kite and twine, so that it can conduct the electric fire freely, you will find it stream out plentifully from the key on the approach of your knuckle. At this key the phial may be charged; and from electric fire thus obtained spirits may be kindled, and all the other electric experiments be performed which are usually done by the help of a rubbed glass globe or tube, and thereby the sameness of the electric matter with that of lightning completely demonstrated.

To connect heavenly forces to the "virtue" that humans had puzzled over since the first sparks were rubbed from amber elevated electricity above the realm of toy science and entertainment. As Philip Dray notes, "Franklin's conclusions demanded that electricity join gravity, light, heat, and meteorology in any account philosophers offered for the majestic workings of nature." Still, half a century after Franklin's kite experiment, at the end of the eighteenth century, in a world illuminated at best by the Argand lamp, the understanding of electricity had hardly advanced any further, hampered in part by the limits of the Leyden jar, which could only bring experimenters so far, since it stored limited energy.

In the late eighteenth century, in Italy, Alessandro Volta challenged Luigi Galvani's conclusion that convulsions in frogs, which Galvani had hung from brass hooks upon an iron trellis, were caused by innate electricity within the animals themselves. Volta argued that the convulsions were caused simply by the contact between the brass and the iron, and he proved his theory by creating the first modern battery, which he described in a letter to the Royal Society in London in 1800:

> I obtain several dozen small round plates or disks of copper, brass, or better of silver, an inch in diameter, more or less; for example coins, and an equal number of plates of tin, or, what is still better, of zinc, of the same shape and size approximately. . . . I prepare besides a sufficiently great number of disks of cardboard, or cloth . . . capable of imbibing and retaining considerable water. . . . I place, generally horizontally, on a table or other base, one of the metallic plates, for example, one of silver; on this first, I then place a second of zinc; on this second, I place a moistened disk; then another plate of silver, followed immediately by another of zinc, to which I can make succeed a moistened disk. I then continue . . . always in the same direction. . . . I continue, I say, to form by many of these sets a column sufficiently high that it may be able to stand upright.

The charge would last for as long as the electrochemical interactions between the liquids and various metals lasted. Volta had created a sustained, continuous flow of electricity. As Park Benjamin, writing in the nineteenth century, noted, Volta's invention "made electricity manageable. He reduced the infinite rapidity of the lightning stroke to the comparatively slow but enormously powerful current, which in the future was destined to carry men's words from one end of the world to the other, and to produce the dazzling light inferior only to the solar ray."

Volta's "pile" immediately intrigued scientists across Europe and America, none more so than Sir Humphry Davy — creator of one of the first miners' safety lamps — who, at the beginning of the nineteenth century, held a post as chemist at the Royal Institution in London. Davy worked at refining Volta's pile and eventually had large batteries built in the basement of the institution's laboratory. He carried out a series of experiments with them, including demonstrations of the first electric lights. In 1802 he succeeded in making a platinum filament glow, if only momentarily, by infusing it with electric current. Then in 1809, with the aid of the largest battery yet — consisting of two thousand pairs of plates — he demonstrated the first lasting electric light, the voltaic arc. He passed a current through a charcoal stick, which served as a conductor of electricity; then he touched another charcoal stick to the first, and a spark jumped from the first to the second. As he pulled them apart, an arc of brilliant blue-white light leapt across the heated air between them. But light wasn't created by the arc alone; the carbons glowed incandescently.

Davy never took the voltaic arc beyond the demonstration stage — an enduring, practical electric light was still many decades away, for considerable problems had to be overcome. Not only did Davy's charcoal electrodes burn quickly and unevenly, but as the carbons burned down and the gap between them widened, the light sputtered, then failed. Scientists had to develop electrodes that would burn slowly and steadily, at a constant distance from each other. The greater challenge, however, lay in producing a more enduring power system than the batteries of the day, and widespread arc lighting would depend on a reliable electric generator, or dynamo, as it was commonly called. That would not arrive until well after 1831, the year Michael Faraday established the principle of electromagnetic induction.

Early arc lights ran on batteries and on small steam-driven generators, but this, as Wolfgang Schivelbush notes, was a step back from gaslight, because there was no possibility of widespread interconnected lighting. Their use was limited to outdoor work yards and lighthouse towers, or for special display and spectacle, as at the coronation of Tsar Alexander II in 1856, when "the city of Moscow was lighted by numbers of electric lamps suspended in the old bell-tower of the Kremlin, a thousand gilded domes glittering in the unearthly radiance, in happy contrast with the quaint arches of the old cathedral close at hand, while the river Moskva was transmuted into a stream of liquid silver."

By the late 1870s, Russian inventor Paul Jablochkoff had made major improvements to the arc lamp. In his design, the carbons, separated by gypsum insulation, stood upright and were set side by side; his "candles" were lit at the top and burned down. Jablochkoff bundled four of them under a glass globe — much more efficient than burning in open air — and devised a regulator so that as one extinguished itself (it could last for about two hours), the next automatically began to burn. The lights ran on the improved generators of the time — by then, Belgian Zénobe Gramme had built a steam-driven dynamo powerful enough to drive a series of streetlights. Jablochkoff's "candles" first lit public halls and department stores, then in 1878 the first arc streetlamps appeared in London and along the Avenue de l'Opéra in Paris, where, being considerably brighter than the traditional gas lamps — perhaps 800 candlepower apiece — they were set about 150 feet apart. Each one replaced up to six gas fixtures.

Until the advent of arcs, street lighting had inched forward, the greasy candles in windows giving way to lanterns, then gaslight, and with each modest improvement — deemed remarkable — life filled up the new space given it in the night. Old light retreated into the far streets and the lesser-known neighborhoods, disregarded and disparaged in relation to the new. But always streetlamps had been light vessels, following the pattern of the streets, and each post cast its own halo that gradually diminished into shadow. People moved in and out of illumination, and streetlamps were an integral part of the streetscape: lamps on the street were in conversation with lamps in homes, cafés, and restaurants; in conversation with the dusk, then the night.

Arc lights fundamentally changed all that. They were exponentially brighter than any previous light — ranging from 500 to 3,000 candlepower. But also the very quality of the light was different. Even under the most efficient oil and gas lamps, as is usual in the dark, the eye saw with its retinal rods. Yet arc lights were so similar to daylight that the human eye worked as it did during the day, using its retinal cones. These lights were so intense that they had to be hung considerably higher than the existing gas and oil streetlamps, above the direct line of human vision, and the light poured down over large areas. The streets no longer appeared as avenues lined with distinct lamps, with "spark lining up with spark." Rather, the light hit off walls and entered houses and was so bright it was claimed that one could see the flies on the walls and read a newspaper streets away from the source. Men and women "suddenly found themselves bathed in a flood of light that was as bright as the sun. One could in fact have believed that the sun had risen. This illusion was so strong that birds, woken out of their sleep began singing. . . . Ladies opened up their umbrellas . . . in order to protect themselves from the rays of this mysterious new sun."

For some the new light stung. Robert Louis Stevenson, upon seeing the first electric arc lights in London, then Paris, wrote:

A new sort of urban star now shines out nightly, horrible, unearthly, obnoxious to the human eye; a lamp for a nightmare! Such a light as this should shine only on murders and public crime, or along the corridors of lunatic asylums, a horror to heighten horror. To look at it only once is to fall in love with gas, which gives a warm domestic radiance fit to eat by. Mankind, you would have thought, might have remained content with what Prometheus stole for them and not gone fishing the profound heaven with kites to catch and domesticate the wildfire of the storm.

But many people didn't dismiss arc lights so quickly. All they'd wanted for so long was *more* light. Now that they might have it in profusion, they had to test the boundaries of brilliance, and small and large municipalities alike pushed forward with public arc light systems. In the United States, inventor Charles Brush, who'd refined an arc lamp at about the same time Jablochkoff was developing his candles, initially illuminated the centers of modest midwestern cities with his systems. The first was Wabash, Indiana, which at the time of Brush's installation had been illuminated by sixty-five gas lamps. Over the courthouse in the middle of town, Brush suspended four 3,000-candlepower arc lamps — the dynamo being driven by a threshing machine engine. On the gloomy, rainy night of March 31, 1880, the arcs were turned on: "Promptly as the courthouse clock struck eight, the thousands of eyes that were turned upward toward the inky darkness over the courthouse saw a shower of sparks emitted from a point above them, small steady spots of light, growing more brilliant until within a few seconds after the first sparks were seen, it was absolutely dazzling. . . . People stood overwhelmed with awe, as if in the presence of the supernatural." Not only were the arc lights dazzling, but Brush offered more light for less money — a double strangeness, to have intensity no longer tied to cost: "The city's 65 gas lamps — deemed inadequate — cost $1,105 per year, not including repairs and maintenance. The Brush lights would light the same as 500 gas lamps equally distributed around the town for less than $800 a year."

The enthusiasm for Wabash's lighting system reached far beyond the city limits, and Brush and other manufacturers of arc lights quickly set up streetlights in Cleveland and other smaller American cities, some of which — Denver, San Jose, Flint, Minneapolis, and Detroit — eventually built towers topped with arc lights in their commercial centers. There was often nothing ornamental about them. For instance, San Jose's tower rose more than two hundred feet above the town. Its six arc lamps could emit a 24,000-candlepower umbrella of light over the commercial district. But the tower, constructed of steel tubes and straddling the intersection of two main thoroughfares, looked as if it belonged along the perimeter of a prison yard. The proponents of such lights saw them as more than a means of gaining security for citizens and increasing commerce. These were new cities establishing themselves in the hinterlands, and they had little history to give them a cosmopolitan air. Historian David Nye asserts that for such towns, "lighting . . . emerged as a glamorous symbol of progress and cultural advancement."

But tower arc lights, Wolfgang Schivelbush suggests, were also more democratic:

> Cities lit in this way were like living Utopias of equality. This was, in fact, one of the main arguments put forward in favour of this type of lighting. The city Council Committee of Flint

(Michigan) justified its decision to introduce a tower lighting system by pointing out "that . . . the light covers the entire space. . . . We claim for it that it may be justly called the poor man's light, for, by reason of its penetrating and far-reaching rays, the suburbs of the city will be equally well lighted with the more central portions . . . and brilliant light will penetrate the most distant parts of the city."

Eventually, so much light proved to be too much. Certainly, bold, bright light would always have its allure — the gaudiness of Times Square, the coronation of a tsar — but it wasn't the light for everyday streets. Shadow life, people found, had its value, too. Municipalities that had embraced the tower arc lights decided to dismantle them and try for something more modest and traditional: lighting that would maintain hours distinct from day — navigable, but also mysterious and a bit hidden; a light that did not dominate them but inhabited the world along with them. They turned away not only from the arc lights but also from the functional appearance of the arc tower. In downtown Minneapolis, where Brush had erected an "electric moon" — a tall post comprising eight arc lamps — the city council "following the practice of many another progressive city, . . . eliminated the ungainly post of iron . . . and substituted therefor[e] upon all of her principal business streets an ornamental lamp post of splendid design with five lights." For added grace, the city installed hanging planters on its new lampposts.

Whereas the smaller cities may have overreached in a large way before drawing back, the life of arc lights in New York City had a slightly different trajectory. New Yorkers were accustomed to relatively bright light; even so, the brilliance of the arcs was startling and brought a bit of unease at first. Brush installed his first arc light system in late 1880, along a street that was illuminated with gaslight. His lampposts weren't commanding towers, but they did stand at twice the height of the gas lamps, a "single light alone being equal to 10 times that of the half-dozen coal gas burners below." The *New York Times* reported that

> the moment the dazzling electric sparks appeared, there was a general turning from the shop windows to the lights. Exclamations of admiration and approval were heard on all sides, together with calculations as to the effect upon the gas companies. . . . Like all electric lights there was a certain intensity about the powerful white rays which, to unaccustomed gazers and persons with weak eyes, somewhat detracted from the pleasure of the illumination. This however will be modified by time and by constant use, while the possibilities of strengthening and softening the rays by the use of ground porcelain or colored glass are such that almost any effect may be produced. As it was last night, the eye, after resting for a time on the dazzling brilliancy of the fierce white jets turned with relief to the mellow golden color which the street lamps and shop windows assumed by contrast.

Over time, porcelain globes did mitigate the light, and in this more modest form, arc lights would illuminate city streets for decades to come. But arc lights had conditioned people to a new level of brightness. By comparison, the gas streetlights that had begun the century, and had once enthralled city dwellers with their soft brilliance and beauty, now appeared dull and ineffectual, much as oil lamps had seemed when gaslight was new. By the end of the nineteenth century, the New York City gaslights, which had in truth become brighter and more reliable over the years, would be perceived as the lights of other days. "Since the electric lamps have become so common in the streets there is scarcely a gas-lighted neighborhood that has not felt that it was being defrauded out of its proper amount of light," an 1898 newspaper article commented. "Occasionally complaints are heard that certain gas lamps are no better than kerosene oil lamps, and the complainants aver that the illuminating power of the gaslights is continually diminishing."

7

INCANDESCENCE

ARC LIGHTS, EVEN WITH SHADES, shone far too intensely to illuminate domestic interiors, and they could not be made less powerful — nineteenth-century scientists would say they were "indivisible." How, then, to make electric illumination intimate enough for the home, equal to the 10 to 20 candlepower of a gaslight fixture? It was a challenge little different in kind from the one Ice Age humans faced when they tamed their hearth fires by fashioning fat-burning lamps such as the ones found at Lascaux.

Working scientists took almost eighty years to "subdivide" electric light. Dozens of experimenters in Germany, England, France, Russia, and the United States worked on the development of an incandescent bulb in the decades after Sir Humphry Davy momentarily made a platinum filament glow at the Royal Institution in 1802, but they encountered seemingly insurmountable problems with the filaments they fashioned out of carbon, a platinum-iridium alloy, or asbestos, which they enclosed in vacuums or sometimes surrounded with nitrogen. Carbon, in all the experiments, quickly destroyed itself. Platinum, which resisted oxidation, tended to fuse when heated to incandescence, and it was expensive. Several experimenters

managed to make a short-lived light from platinum in an evacuated bulb, the most lasting being William Grove's, which in
1840 illuminated an English theater for the length of a performance, though the light was dim and costly.

By the 1870s, all the problems scientists had encountered
with incandescence throughout the century persisted, and the
field was crowded with those who were still trying to adequately evacuate bulbs and craft enduring filaments. Among
them were Hiram Maxim, Moses Farmer, William Sawyer,
and Albon Man in the United States and St. George Lane-Fox
and Joseph Swan in England. Swan had been attempting to
make a filament lamp for thirty years, and the transactions of
the December 1878 meeting of the Newcastle Chemical Society note that he "described an experiment he had recently performed on the production of light, by passing a current of
electricity through a slender rod of carbon enclosed in an exhausted globe. . . . The rod became heated to such an intense degree as to cause it to glow with great splendour." Still,
Swan's light was only momentary, and the glass bulb quickly
became coated with soot.

Thomas Edison joined the fray in the late summer of 1878.
"It was all before me," he was to later say. "I saw the thing had
not gone so far but that I had a chance. I saw that what had
been done had never been made practically useful. The intense
light had not been subdivided so that it could be brought into
private houses." Edison knew he not only had to find a durable
material and an ideal shape for the filament, but he also had to
produce an adequate insulating material and figure out how
to quickly, efficiently, and completely evacuate the air from a
glass bulb. He had to create an effective delivery system for
electricity — which meant developing workable switches and
wiring — and an efficient dynamo. The dynamo was a particular challenge (earlier electric devices, such as telegraphs and

telephones, could run on batteries), and he found the solution to this problem during a trip to engineer William Wallace's factory in Connecticut, where Wallace manufactured both arc lamps and dynamos. There Edison encountered Wallace's "telemachon," a dynamo powerful enough to illuminate eight arc lamps simultaneously. The machine utterly inspired him. A *New York Sun* reporter who'd accompanied Edison on the trip noted that Edison "ran from the instrument to the lights and from the lights back to the instrument. He sprawled over a table with the SIMPLICITY OF A CHILD, and made all kinds of calculations. He estimated the power of the instrument and of the lights, the probable loss of power in transmission, the amount of coal the instrument would save in a day, a week, a month, a year, and the result of such saving on manufacturing." Edison himself remarked, "Now that I have a machine (Wallace's) to make the electricity, I can experiment as much as I please."

The challenges Edison faced were not only technical. Everything concerning the system needed to be cost-effective, practical enough for general use, and familiar enough to old forms to be easily adopted by the public, which meant being cleaner, more efficient, and more economical than the dominant late-nineteenth-century type of urban indoor lighting: gas. As Edison's mathematician, Francis Upton, was to write, "A mistaken idea has been afloat that this new light was intended to be a rival of the sun, rather than what it really is, — a rival of gas." But Edison saw gas as a model as well as a rival. In early installations, he threaded his wires through existing household gas lines and adapted existing gas fixtures for use with his light. He developed a way to determine the cost of electricity per household based on the gas meters of the time, and he envisioned a system, as for gas, linked to adjoining sites all fed by a central station, which meant it would depend on density — a high volume of use in a small area — for its cost-effectiveness.

From the beginning, Edison understood his system to be an urban one, and he — backed by New York money and followed most closely by the New York press — saw New York City as the foremost testing ground for his work: he planned to install his first commercial central lighting system in Manhattan. In clear weather, he could see the city on the horizon, thirty or so miles away from his laboratory in Menlo Park, New Jersey, where he carried out his first electric light experiments. Menlo Park had been a failed real estate venture — no more than a handful of modest houses on a hill and a whistle stop on the Pennsylvania Railroad line — when Edison, feeling cramped and crowded in his Newark laboratory, began looking for a new site. At Menlo Park, he found isolation and plenty of inexpensive land on which to build his compound.

At first glance, you feel a sense of seclusion and of quiet containment when you encounter R. F. Outcault's painting of the laboratory and its surroundings in the winter of 1880–1881. It might be a farm settled into its yearly sleep, for it is laid out nearly the same. The compound, standing squarely in the midst of open fields, is surrounded by a picket fence. A road running alongside disappears into woodlands at the rim of the horizon. The library/office in the foreground resembles a modest two-story clapboard house, and the clapboard laboratory behind it, other than having small porches on each floor, could be an elongated barn. Off to the side is a shed with a ladder propped against it.

But it was, for all its traditional appearance and apparent modesty, the largest private laboratory in the United States. The picture also includes a red brick machine shop with a smokestack at the rear of the site, telegraph poles strung with wires across the near field, and a train parked at the whistle stop on the far side of the road. What went on there was new and bewildering to any outside observer. "When I was a boy," David Trumbull Marshall remembers,

being a boy, and consequently of no account, I was allowed to roam through the Laboratory at Menlo Park. . . . I remember seeing the tall lean Mr. Lawson firing the furnaces for carbonizing the lamp filaments. . . . I remember seeing men in the yard outside the Machine Shop . . . laboriously winding copper wires with roller bandages and slushing the whole with asphalt. . . . I remember going into the little blacksmith shop on the south side of the Laboratory inclosure and there finding a blacksmith [who] was making something out of copper and [he] told me that "it was a very particular thing to be done." . . . I remember the small shed next [to] the blacksmith shop in which there were a number of kerosene lamps burning, the flame turned up purposely so that the flame would smoke and deposit soot. . . . I remember the day Alfred Moss and I discovered the rubbish heap. . . . We thought we had struck a gold mine. Pieces of insulated copper wire, pieces of glass tubing, pieces of brass and the other thousand and one things that drop on the floor and are swept up and thrown out.

Within, Edison's "invention factory" was peopled with blacksmiths, electricians, mechanics, machinists, model makers, a glassblower, and a mathematician. "His iron ideas, in tangled shapes, are scattered and piled everywhere; turning lathes are thickly set on the floor and the room is filled with the screech of tortured metal," one reporter wrote. "Upstairs . . . is walled with shelves of bottles like an apothecary shop, thousands of bottles of all sizes and colors. . . . On benches and tables are batteries of all descriptions, microscopes, magnifying glasses, crucibles, retorts, an ash-covered forge, and all the apparatus of a chemist."

As the *New York Herald* reported, work there went on all through the night:

> At six o'clock in the evening the machinists and electricians assemble in the laboratory. Edison is already present, attired in a suit of blue flannel, with hair uncombed and straggling over his eyes, a silk handkerchief around his neck, his hands and face somewhat begrimed. . . . The hum of machinery drowns all other sounds and each man is at his particular post. Some are drawing out curiously shaped wires so delicate that it would seem an unwary touch would demolish them. Others are vigorously filing on queer looking pieces of brass; others are adjusting little globular shaped contrivances before them. Every man seems to be engaged at something different from that occupying the attention of his fellow workman.

Edison focused a good deal of attention on the search for the best material for the filament, having concluded that the filament for the bulb had to be constructed of high-resistance material. "The more resistance your lamp offers to the passage of the current," he explained, "the more light you can obtain with a given current." In the course of many months, his crew tried and abandoned materials — carbon, platinum, silicon, boron — and then returned to carbon, which offered high resistance, although it was difficult to stabilize. They carbonized fishing line, rosewood, hickory, spruce, coconut fibers, and countless other substances. They shaped filaments as boxes, spirals, circles, horseshoes, and fanciful sprouts and curlicues, recording every experiment in a series of notebooks. To glance at even a few of the entries is to begin to comprehend the breadth and detail of their attempts:

(April 29) Wood loop cut from the thin worked holly milled by Force and cut after manner and in same former used for cardboard, carbonized by Van Cleve, were measured and put in lamps ready for pump, resistance 125 and 194 ohms.

(May 14) Carbonization. Several moulds of Bast fibers were carefully prepared and formed around wood for carbonization, but the wood proved very detrimental, every one having been broken in the moulds during the process. Van Cleve is preparing some more for trial.

(May 20) Carbonization. Van Cleve carbonized three moulds of bent wooden loops by securing the strips in slotted nickel plates; he got them out very nicely and in good shape. Bast fiber. Four of the Bast fiber lamps were measured and tested with current of 103 volts[;]they gave 30 to 32 candles and about six per horse power. They were connected to main wires in Laboratory and during the first hours three of them broke in the clamps and glass but the fiber in each instance remained in the globe unbroken. Showing the fiber to make strong carbon but difficult to form good contact with.

On October 22, 1879, Charles Batchelor, Edison's most trusted associate, placed a horseshoe-shaped filament of carbonized cotton thread in an evacuated handblown glass bulb and attached it to a series of batteries. The bulb began to glow at 1:30 A.M. and glowed through the rest of the night and the following morning. At 3:00 in the afternoon, he added more battery cells for additional power, and Batchelor noted that the bulb became as bright as three gas jets of the time, or four kerosene lamps: the brilliance of about thirty candles. An hour later, in the waning of a late-fall afternoon, the glass bulb cracked. It had burned for more than fourteen hours.

Everyone working at Menlo Park knew that all aspects of the system needed further improvement before the light would be commercially viable — the dynamo, the switches, the bulb, and the filament, for which they would eventually turn to bamboo. Still, by December Edison was able to display his system to his financial backers. At the same time, he showed it to his friend, reporter Edwin Fox, who took notes for a long article intended to be printed after the official public display of the system, which was planned for New Year's Eve. But when news leaks began appearing in other papers, Fox's newspaper, the *New York Herald*, decided to print a full-page story on December 21. "Edison's electric light, incredible as it may appear, is produced from a tiny strip of paper [actually carbonized thread] that a breath would blow away," wrote Fox. "Through this little strip of paper passes an electric current, and the result is a bright, beautiful light."

Upon publication of the story, thousands traveled to see Edison's invention for themselves. Wealthy New Yorkers, accustomed to the glow of gas lamps, arrived from their city in horse-drawn carriages. Others traveled by trains that steamed through the short, cold afternoons. Farmers who lived solely by the light of kerosene lamps rode in from the dark countryside, hauling wagonloads of children riding atop bales of hay. In the Menlo Park lab, they jostled one another to look at the light of the future, which would arrive for the wealthy in a handful of years. The farmers — and the farmers' children — might never live to see it in their homes, but perhaps for the moment, before its history unfolded, they were all equal in their wonder. Here was a little click that meant light was contained in a glass vacuum and need never again be linked with a flame or coaxed forth and adjusted; light that did not waver, tip, drip, stink, or consume oxygen and would not spontaneously ignite cloth dust in factories or hay in the mow. A child could be left alone with it.

Surely, a measure of its beauty and brilliance was linked to the Menlo Park setting, a place that must have been both familiar and strange: with its blacksmiths and glass blowers, but also its mathematicians and electricians, notebooks in hand. And a place remote and apart in the deep heart of winter, the time when light holds its greatest meaning. The contrast between the greater dark and the glowing "strip of paper" could only have reinforced how those present were witness to something they had never imagined, something that would change the quality of shadows as well as the quality of light, and change the atmosphere of their household nights.

The crowds continued to arrive by day and by night, so many that after a few days, Edison was forced to close his lab to them. But he kept the lamps burning so that those who came could view them from the grounds. When he opened up the laboratory again on New Year's Eve to officially display his lighting system, thousands more arrived at Menlo Park to see the twenty-five lamps in the lab, the eight in the counting room and office, the twenty on the street and in neighboring houses. The *New York Herald* reported:

> The light was subjected to a variety of tests. Among others the inventor placed one of the electric lamps in a glass jar filled with water and turned on the current, [and] the little horseshoe filament when thus submerged burned with the same bright steady illumination as it did in the air. . . . Another test was turning the electric current on and off on one of the lamps with great rapidity and as many times as it was calculated the light would be turned on and off in actual house illuminations in a period of thirty years, and no perceptible variation either in the brilliancy, steadiness or durability of the lamp occurred.

By the following winter, Edison, having successfully expanded his electric system around Menlo Park by means of underground conduits, shifted his operations to Pearl Street in Manhattan, with the intention of developing a practical and workable central station that would deliver power to the surrounding neighborhood. During the several years it took him to complete the Pearl Street station, he installed incandescent lights as isolated direct current (DC) systems, first on the cruise ship *Columbia* and then in factories throughout the country. Manufacturers whose businesses were prone to fire, such as sugar refineries, were immediately interested in Edison's system — this light without flame or sparks — as were textile manufacturers, lithographers, and paint manufacturers. In their factories, better light would make quality work much easier to accomplish.

A reporter who visited the Merrimack Mills in Lowell, Massachusetts, where Edison had installed a system in early 1882, described the palpable difference the light made:

> Standing at one end of the room and glancing down the long rows of looms, each with its own little light placed three feet above the fabric being woven, one is first struck by the agreeable quality of the light, and next by its perfect steadiness. . . . The absence of heat is another valuable quality. . . . The temperature of the room, which would be raised ten or twelve degrees by the lighting of the gas, is not influenced by the 262 electric lights. Upon approaching the loom and examining the work in process, it appears that every thread, every line of pattern in fancy plaid goods, is remarkably clear and distinct; imperfections are quickly noticed and as quickly remedied, and it would seem that the operatives could desire no more perfect light.

Machines are most economical when they run all the time, and electric lights proved so efficient that their use extended the workday, which had been gradually losing its dependence on natural daylight since the introduction of the mechanical clock in the sixteenth century. Edison's success helped to fully establish the three-shift day and the final erasure of natural time in the factory.

A few months after Edison outfitted the Merrimack Mills, he wired the first private home for incandescent light, banker and financier J. P. Morgan's Madison Avenue brownstone. What was a boon to business was still overwhelmingly complicated for a house and — costing as much as 28 cents per kilowatt hour — only for the very wealthy. "It was a great deal of trouble to install it," recalled Morgan's son-in-law and biographer, Herbert Satterlee. "A cellar was dug underneath the stable . . . and there the little steam engine and boiler for operating the generator were set up. . . . The gas fixtures in the house were wired, so that there was one electric light bulb substituted for a burner in each fixture. Of course there were frequent short circuits and many breakdowns on the part of the generating plant." Edison's engine, fueled by coal, belched. It was noisy. It sent up foul smoke and fumes. The neighbors complained that their houses shuddered when the boiler started up in the afternoon. It wasn't self-starting or self-maintaining. Sometimes, the house was plunged into darkness when a generator broke down or the wires short-circuited. In addition, the generator

had to be run by an expert engineer who came on duty at three P.M. and got up steam, so that any time after four o'clock on a winter's afternoon the lights could be turned on. This man went off duty at 11 P.M. It was natural that the family should often forget to watch the clock, and while visitors were still in the house, or possibly a game of cards was going on, the lights would die down and go out. If they wanted to give a party, a special arrangement had to be made to keep the engineer on duty after hours.

Morgan proved to be a patient customer: even after a renovation in which an electric cord running under the rug in his study started a fire that destroyed the room, he stood by the system. But not every wealthy client of Edison's was so intrepid or comfortable with the industrial obviousness of it. Gaslight involved no boilers or burners in the home, removed as it was from its grimy source of power by miles of pipe. The wife of railroad magnate William H. Vanderbilt refused to use the electricity in her newly wired home because she feared living over the boiler.

Meanwhile, Edison's plans for the central power station on Pearl Street in Manhattan made slow progress, in part because of the laborious task of digging subways to lay underground wires for the system. Edison insisted on buried wires, not only to follow the example of gas lines but also for one of the same reasons he favored low-voltage direct current over higher-voltage alternating current (AC): safety. In the commercial districts of Manhattan, even before the advent of electric arc streetlights in 1880, numerous small companies had strung wires for telegraphs, telephones, alarms, and stock tickers along poles down city streets. Wires sagged across avenues, weighed down top-heavy crossbars, and were tautly anchored down to the sides of buildings. Each company was responsible for the upkeep of its own lines, and not uncommonly, through neglect or storm damage, loose wires sagged or dangled from poles. Companies went out of business but failed to take down their wires, which simply deteriorated in place. The lines, at first, were problematic but not deadly, as most services ran their operations off batteries. But as historian Jill Jonnes observes,

> All that changed with the coming of the new outdoor arc lighting.... The extremely high voltage alternating current required

to operate these lights — as high as 3,500 volts — made their outdoor wires truly perilous. The Brush Electric Company . . . built three central power stations and transmitted its high-power electricity — typically 2,000 to 3,000 volts — on wires strung among the existing low-voltage tangle. Edison wanted nothing to do with these mangled nests of live and abandoned wires.

So he worked on his subways, while a competing, unorganized lighting market grew throughout the city. Arc light companies illuminated streets, large public buildings, theaters, and hotel lobbies, while incandescent light companies built isolated systems for the interiors of buildings. Hiram Maxim, for instance, had successfully wired incandescent lights in the Mercantile Safe Deposit Company in Manhattan by late 1880. Gas companies responded to electric light by attempting to create more powerful, efficient gas lamps, an effort that would culminate in the development of the Welsbach lamp, which consisted of a burner — essentially a Bunsen burner — surrounded by a mantle composed of finely woven cotton fabric that had been impregnated with a solution of oxides and then dried. Although the burner consumed oxygen and overheated rooms just like traditional gas burners, the mantle, first advertised in 1890, glowed incandescently. It gave off impressive light, although it was also fragile. The lamp was marketed as the "electric light without the electricity."

Finally, Edison completed enough of his system by the summer of 1882 to partially light the neighborhood around Pearl Street, which included the offices of the *New York Times*. On September 4 of that year, he turned on his system, and those working in the newspaper office seemed particularly grateful:

It was a light that a man could sit down under and write for hours without the consciousness of having any artificial light

about him. . . . The light was soft, mellow, and grateful to the
eye, and it seemed almost like writing by daylight to have a
light without a particle of flicker and with scarcely any heat to
make the head ache. The electric lamps in THE TIMES Building
were as thoroughly tested . . . as any light could be tested in a
single evening, and tested by men who have battered their eyes
sufficiently by years of night work to know the good and bad
points of a lamp, and the decision was unanimously in favor of
the Edison electric lamp as against gas.

Those on the street at first hardly noticed the modest light.
The *New York Herald* reported:

In the stores and business places throughout the lower quarters
of the city there was a strange glow last night. The dim flicker
of gas, often subdued and debilitated by grim and uncleanly
globes, was supplanted by a steady glare, bright and mellow,
which illuminated interiors and shone through windows fixed
and unwavering. From the outer darkness these points of light
looked like drops of flame suspended from jets and ready to fall
at every moment. Many scurrying by in preoccupation of the
moment failed to see them, but the attention of those who
chanced to glance that way was at once arrested. . . . The test
was fairly stood and the luminous horseshoes did their work
well.

Direct current also did its work well in sending low voltages
over short distances, but it had limitations. First, after a half
mile or so, the current quickly diminished and could not be
bolstered without costly outlay for thick copper wiring. Second, although direct current could adequately serve electric
light customers by delivering a steady 110 volts, more powerful currents to run motors couldn't travel over the same lines.
In addition to these intrinsic problems, negotiations for cen-

tral stations were often complex, since many parties with differing interests would have to come to an agreement. For all of the initial success of Pearl Street, by the end of 1884 Edison had built only eighteen central stations (compared with hundreds of isolated systems that electrified individual homes and businesses).

The true threat to Edison's system proved to be alternating current stations, which sent high-voltage current over wires to transformers that stepped down the power to a lower voltage before delivering it to individual homes and businesses. Alternating current could accommodate different voltages, so the system could power both lights and motors, and the stations could send, via thin copper wires, a steady, strong power supply farther than the half-mile radius of direct current systems. Alternating current systems could expand outward as growth warranted.

Perhaps no one understood the advantages of alternating current more than George Westinghouse, who formed Westinghouse Electric Company in Pittsburgh in 1886 and subsequently contracted with inventor Nikola Tesla to help develop alternating current systems for his company. Tesla: tall and lean, possessing intense blue eyes, hypersensitive to the sun and to the experience of passing under a bridge, which caused pressure on his skull. "I would get a fever from looking at a peach, and if a piece of camphor was anywhere in the house, it caused me the keenest discomfort," he once said. "When a word was spoken to me, the image of the object it designated would present itself vividly to my vision and sometimes I was quite unable to distinguish whether what I saw was tangible or not." He seemed to live in a fever and quieted his mind by counting his own footsteps during his frequent walks. However much this fever might have been a burden, it was also essential to his creations. He could build machines entirely in his head, down to the smallest details, understand how they worked, and know how they needed to be improved. He could revise them without ever setting them down on paper or making a model.

After immigrating to the United States in his twenties, Tesla worked briefly for Edison, who never seemed to truly acknowledge his genius and refused to pay him a promised bonus, after which Tesla left Edison's employ. But even before their final falling-out, Tesla felt hampered by Edison's fidelity to direct current. When he brought up the subject of alternating current, Edison snapped, "Spare me that nonsense. It's dangerous. We're set up for direct current in America. People like it, and it's all I'll ever fool with."

Edison publicly condemned alternating current. "It will never be free from danger," he declared, while also claiming it was unreliable and unsuitable for central station systems. He took to calling alternating current "the executioner's current," and after promoting a series of high-profile electrocutions of animals — a dog, a calf, and finally an elephant — to prove its fatal power, he publicly supported its use for the first electric chair. The competition between Westinghouse's alternating current and Edison's direct current played out publicly and bitterly in what came to be known as the War of the Currents.

Edison at first seemed to have public anxiety on his side after a series of events in New York City reinforced the dangers of high-voltage wires. First, in the winter of 1888, a blizzard crippled the city: "The wind at times seemed to make the entire circuit of the compass, and men and women were whirled about it like so many dolls. The snow was sharp and dry, and . . . it cut like so many pieces of glass. It clung to whiskers and froze . . . until the hair on men's faces [was] transformed into glistening miniature icebergs." During the storm, electric lines throughout the city came down. "Poles, with their long arms

laden with wires and cables, were wrenched and twisted mer-
cilessly by the wind. Roof fixtures, with their tangled masses
of twisted and broken wires, met the eye on all sides, and the
loose ends, lashed by the wind, whistled through the air like
whipcord. . . . The breaking of the telegraph, telephone, and
electric light wires, with the danger to vehicles and pedestrians attendant thereon, was added to by the danger of falling
poles."

The devastation alarmed both citizens and officials, and the
alarm was compounded by a subsequent series of "deaths by
wire" in the following months, including that of a young boy
who was electrocuted after he playfully jumped up to touch a
dangling wire. When, in the fall of 1889, a telegraph company
employee was killed as he worked on the lines, his brutal death
was witnessed by a crowd of New Yorkers: "The man appeared
to be all on fire. Blue flames issued from his mouth and nos-
trils and sparks flew about his feet." A public outcry ensued,
and the mayor ordered several light companies, which illuminated three-quarters of the city below Fifty-ninth Street, to
extinguish their streetlights and repair their lines before lighting up again. Darkness fell over much of a city accustomed to
light. The *New York Times* reported that the

aspect of the city was decidedly provincial. . . . In the vicinity of Union and Madison squares, City Hall Park, and other
open spaces the view was particularly cheerless and depressing.
Thoroughfares like Broadway, Fifth, Madison, and Seventh avenues looked by contrast like endless tunnels of gloom. . . . The
Edison system was working as usual in all the Broadway and avenue stores and in all public places through the central section
of the city, where its subways are laid. . . . Orders were at once
sent out to all police stations in the darkened district that a double patrol force should be sent out and patrolmen given special

instructions to use extra vigilance while on post in guarding life and property from footpads.

Westinghouse countered Edison and sought to assure the public by insisting on the safety of well-constructed lines: "As to the accidents from electric currents," he wrote, "the records of deaths in the city of New York show that there were killed by street-cars during the year 1888, 64 persons; by omnibuses and wagons, 55; and by illuminating gas, 23; making the number killed by the electric current (5) insignificant compared with the deaths of individuals from any of the other causes named."

However dangerous it appeared to be, versatile alternating current was also the ideal current for a rapidly expanding nation and its economy. Although electricity was still almost fully aligned with light in most minds, and the growing number of companies that produced and sold electricity were still called "light companies" rather than "power companies," the mechanical uses of electricity had begun to emerge: electricity began to drive all kinds of devices and machines for factories and households. By 1891 alternating current systems had begun to gain favor; there were almost five times as many alternating current stations in the country as there were direct current stations. Then George Westinghouse outmaneuvered Edison's General Electric Company for the major contract to supply electricity to Chicago's World's Columbian Exposition of 1893, meant to celebrate the four hundredth anniversary of Columbus's voyage to America. Tesla's polyphase generators would power the greatest ode to electric light the world had yet seen, and the momentum alternating current gained from the exposition would consign direct current to the past.

8

OVERWHELMING BRILLIANCE: THE WHITE CITY

Electricity is the half of an American.

—HUBERT HOWE BANCROFT,
The Book of the Fair

THE WORLD'S COLUMBIAN EXPOSITION of 1893 —
the largest world's fair up to that time — sprang from
the most unpromising stretch of land: "a marsh when
work upon it was begun, a sopping combination of low lands,
water, and hummocks," noted one observer. Another called it
"a treacherous morass, liable to frequent overflow . . . bearing
oaks and gums of such stunted habit and unshapely form as
to add forlornness to the landscape." Over three years' time,
thousands of men downed the trees; dredged out the muck and
hauled it away in wheelbarrows; reshaped more than six hundred acres along Lake Michigan — six miles from downtown
Chicago — into promontories and islands; and constructed
viaducts, bridges, pathways, and paved boulevards. Countless
more skilled workers and laborers — using more than eighteen thousand tons of iron and steel — framed fourteen massive
structures around a broad lagoon and plaza to create the Court
of Honor, the centerpiece of the fair.

Although a different architect designed each building in the court, chief planner Daniel Burnham required all of the buildings to be adorned with neoclassical arches, towers, and pinnacles; all of the cornices to be set sixty feet above the ground; and all of the edifices to be painted white — the shade, one observer noted, "of darkened ivory or slightly smoked meerschaum." Such unified architecture, Burnham imagined, would create an exposition reminiscent of Venice, without the grime, raw sewage, or ruins. He even imported sixty gondolas from Italy to carry passengers along the waterways. The court came to be known as the White City, in part for the way its pale edifices gleamed in the prairie night.

Never had there been so much light in one place — and it was all electric: 200,000 incandescent bulbs traced the edges of the edifices, and countless more lit the interiors of the massive exhibition halls; 6,000 arc lights on twelve-foot-high posts lined the paths and walkways. That light glinted in the lagoons and bounced off the fountain waters; it glittered in the wakes of the gondolas and the currents of Lake Michigan. Such brilliance seemed all the more miraculous because there were no leaning poles and sagging wires, nothing obvious carrying the current: so as not to mar the beauty and unity of the buildings, the wires ran underground.

Colored lights shone as well. From the rooftops, searchlights fashioned with blue, green, red, and violet slides swept the city and waterways; colored bulbs illuminated water fountains "so bewildering no eye can find the loveliest, their vagaries of motion so entrancing no heart can keep its steady beating." Every night, fireworks went off from different locations. "There would be a dozen or more rockets sent up all at once, and they would all explode together, almost filling the air with red, blue, and green stars, which floated . . . for a moment, and then dropped slowly into the water," remembered one fairgoer. The incandescent bulbs, the arc lamps, the search-lights, the fireworks — separately each would have astonished nineteenth-century eyes; together, they overwhelmed. "It is the part that each one plays in the general effect," wrote one commentator, "all contributing to give this wondrous display the aspect of a veritable fairyland, to raise it, for the moment, almost beyond the realm of matter."

Night at the fair had come a long way since London's Crystal Palace, or Great Exhibition of 1851, which had closed at dusk. Not until the 1867 Paris Exposition did a world's fair stay open at night. There gas and oil lamps "were used lavishly[,]. . . music and theatricals were supplied of satisfactory character and quality, restaurants and cafés were kept open, and the exposition generally was given as gay and festive an air as possible. The prodigal expenditure of time, money and labor were without avail, however, and the effort to force attendance after dark was a signal failure, solely because of the insufficient light, and the refusal of the people to be entertained in the dark." Only in the 1880s did evenings at fairs and expositions begin to succeed. Most notably, at the Paris Exposition of 1889, the grounds were officially illuminated with more than a thousand arc lights and almost nine thousand incandescent bulbs (in addition to private displays).

The White City not only had more lights than the 1889 exposition; it had more lights than any real city in the country. Every day, the lights at the exposition consumed three times the electricity used to illuminate nearby Chicago. And the fair required electricity for mechanical power as well: a moving sidewalk equipped with chairs transported people who arrived by boat from Lake Michigan to the heart of the fairgrounds; electric boats, along with the gondolas, ferried people across the manmade lake — lined with statuary and dotted with fountains — at the city's center; and the world's first Ferris wheel carried passengers seated in Pullman cars 264 feet in the air, giving them a kaleidoscopic view of the city, Lake Michigan, and the Illinois, Indiana, and Michigan countryside before it brought them back to earth.

If Chicagoans, who had already grown accustomed to electric light and gaslight, were stunned by the "brilliance almost too dazzling for the human eye to rest upon," how must it have seemed to the many visitors from small settlements and farms along the Mississippi Valley and in the surrounding states who'd left homes that were illuminated only with oil lamps and candles? To those from rural places everywhere? As one young girl, newly arrived from Poland, exclaimed, "Having seen nothing but kerosene lamps for illumination, this was like getting a sudden vision of Heaven." Country visitors knew that the future lay in the cities — the young had been leaving the farms for decades, and rural life, based on the self-sufficiency of the family, had ceased to be typical. To them, the fair might have been not only dazzling but also consoling in its stark contrast to Chicago or any other late-nineteenth-century American city, for the White City — full of oddity, irony, brilliance, grace, and absurdity — was also a dream city, one without the burden of reality: a city without factories or tenements, skyscrapers, stockyards, slaughterhouses, trash heaps, coal ash, or tax collectors. Its furnaces ran on oil piped in from forty miles away, and the tenders wore white uniforms. What trash the visitors scattered about the grounds was picked up every night and carted away.

Chicago, with a population of more than a million, was the American city of the moment, having grown and flourished, observed architect Louis Sullivan, "by virtue of pressure from without — the pressure of forest, field and plain, the mines of copper, iron and coal, and the human pressure of those who crowded in upon it from all sides seeking fortune." Along with its stockyards, train yards, smokestacks, and factories, it could brag of having two dozen skyscrapers — more than any other city at the time — as well as three dozen railroads and hundreds of millionaires. Advertising splashed across the sides of streetcars and loomed on large billboards. "Chicago, one might say, was after all only a Newer York," suggested writer and editor William Dean Howells, "an ultimated Manhattan, the realized ideal of that largeness, loudness and fastness, which New York has persuaded the Americans is metropolitan." Electric wires cluttered the air above the streets. The elevated railways clanged and screeched. Grime and soot settled upon the city's countless poor and working poor, their broken-down tenements, and the red-light district. "'Undisciplined' — that is the word for Chicago," proclaimed H. G. Wells, "a scrambling, ill-mannered, undignified, unintelligent development of material resources."

Strange to think that much of it had risen out of the ashes of its infamous and devastating fire in 1871. Stranger still to consider that sixty years before the exposition, at a time when thousands of gaslights already lined the streets of London and Paris, Chicago was a French and Indian trading village of fewer than four thousand residents, its homes and shops illuminated with tallow lamps and candles. The area had been home to the Prairie Potawatomis, known as the People of the Place of Fire for the way they set the country alight to burn off young trees and old grasses so as to keep the prairie vigorous for game, a world where people kept alive even the smallest flame brought to life from the friction between hardwood and softwood.

Perhaps they husbanded fire in the manner of the Blackfeet, who had once inhabited the country west of Illinois. Around the time of the White City, naturalist George Grinnell wrote of the them:

Within the memory of men now living . . . fire used to be carried from place to place in a "fire horn." This was a buffalo horn slung by a string over the shoulder like a powder horn. The horn was lined with moist, rotten wood, and the open end had a wooden stopper or plug fitted to it. On leaving camp in the morning, the man who carried the horn took from the fire a small live coal and put it in the horn, and on this coal placed a piece of punk [a fungus that grows on birch trees, which the Blackfeet gathered and dried], and then plugged up the horn with the stopper. The punk smouldered in this almost air-tight chamber, and, in the course of two or three hours, the man looked at it, and if it was nearly consumed, put another piece of punk in the horn. The first young men who reached the appointed camping ground would gather two or three large piles of wood in different places, and as soon as some one who carried a fire horn reached camp, he turned out his spark at one of these piles of wood, and a little blowing and nursing gave a blaze which started the fire. The other fires were kindled from this first one, and when the women reached camp and had put the lodges up, they went to these fires, and got coals with which to start those in their lodges. The custom of borrowing coals persisted up to the last days of the buffalo, and indeed may even be noticed still.

At the World's Columbian Exposition, the Native American exhibits were installed in or near the Anthropological Building. According to historian Robert Rydell, "The Native Americans who participated in the exhibits . . . were the victims of a torrent of abuse and ridicule. With Wounded Knee only three years removed, the Indians were regarded as apocalyptic threats to the values embodied in the White City." The verbal threats, perhaps, weren't the worst they had to endure. As it so happened, Rydell notes, "several of the exhibits of Dakota, Sioux, Navajos, Apaches, and various northwestern tribes were on or near the Midway Plaisance, which immediately degraded them."

The Midway Plaisance, a mile-long entertainment district, led up to the entrance of the White City proper. The cultures deemed by the organizers as "barbarous and semi-civilized" were jumbled there along with food concessions and the Ferris wheel: a Moorish mosque, a Tunisian village, an Egyptian temple, a bazaar from India selling Benares brassware and inlaid metalwork, the huts of South Sea Islanders, a settlement of Laplanders complete with reindeer that pulled sleds around a circus ring. The official history of the exposition notes: "Here was an opportunity to see these people of every hue, clad in outlandish garb, living in curious habitations, and plying their unfamiliar trades and arts with incomprehensible dexterity. . . . There were three thousand of these denizens of the Midway gathered from all quarters of the earth."

If, in the future, the honky-tonk sideshows and game booths of midways would be most garishly lit, in 1893 this very first midway claimed a smaller portion of electric light than other areas of the fair, although that didn't stop it from being enormously popular in the evening. As visitors arriving from Chicago walked along the mall toward the entrance to the White City, they could sample chapati and yogurt, Cracker Jack, stuffed cabbage, hamburgers, or steamed clams while they watched boxing matches, donkey races, beauty contests, camel drivers, belly dancers, and sword fighting in a street typical of Algiers. They could listen to a German brass band, Sumatran gong players, Chinese cymbalists, or Dahomean tom-tom players.

The Dahomey village housed sixty-nine people, "of whom twenty-one were Amazon warriors," notes the official history. "Sight-seers . . . were fascinated with the savagery of the fetich war dance performed by the Amazons." This exhibit was particularly galling to African American writer and lecturer Frederick Douglass, an ex-slave: "As if to shame the Negro," he wrote, "the Dahomians are also here to exhibit the Negro as a repulsive savage. . . . It must be admitted that, to out-ward seeming, the colored people of the United States have lost ground and have met with increased and galling resistance since the war of the rebellion." Almost thirty years after the end of the Civil War, the black population of the country stood at more than 7.5 million, yet not one black person had been included in the planning committee for the exposition. "When it was ascertained that the seals and glaciers of Alaska had been overlooked in the appointment of National Commissioners, it was a comparatively easy task for the President to manipulate matters so that he could give the far away land a representa-tive," observed Ferdinand L. Barnett, editor of Chicago's first black newspaper. "It was entirely different, however, with the colored people. When the fact was laid before the President that they had been ignored and were entirely unrepresented, he found his hands tied."

Not only had black people no representation on the plan-ning committee, but they also had almost no formal presence at the fair. The White City housed more than sixty-five thou-sand exhibits, which seemed to one observer to be "the con-tents of a great dry goods store mixed up with the contents of museums." It included a Japanese teahouse, the dungeons of the Inquisition, and the electric chair; sea anemones, devilfish, sharks, catfish, and perch; Bach's clavichord, Mozart's spinet, and Beethoven's grand piano; almost every known fruit and vegetable seed; examples of pests that afflicted crops and pesti-cides used to counter them; more than a hundred exhibits on tobacco and more than another hundred on nuts; a Statue of Liberty carved out of salt; a thirty-five-foot tower of navel oranges — the oranges changed every few weeks — topped by a stuffed eagle; a Liberty Bell made out of wheat, oats, and rye; a map of the United States made out of pickles; and a 22,000-pound mass of cheese encased in iron. Within that glut of variousness, African Americans could claim only several exhibits by black colleges; a painting by George Washington Carver; Edmonia Lewis's sculpture of Hiawatha; and "Aunt Jemima," portrayed by a former slave who wore a red bandana and flipped pancakes outside the R. T. Davis Milling Company booth.

To counter and protest the lack of a dignified presence for blacks, Douglass, antilynching activist Ida B. Wells, Irvine Garland Penn, and Ferdinand L. Barnett published a pamphlet, *The Reason Why the Colored American Is Not in the World's Columbian Exposition*, which detailed the successes of blacks, the colleges they had established, and the inroads they'd made in medicine, law, and the arts. "We earnestly desired to show some results of our first thirty years of acknowledged manhood and womanhood," Douglass wrote in his introduction. "Wherein we have failed, it has been not our fault but our misfortune, and it is sincerely hoped that this brief story, not only of our successes, but of [our] trials and failures, our hopes and disappointments will relieve us of the charge of indifference and indolence. . . . And hence we send forth this volume to be read of all men."

The struggle of African Americans for a presence in the White City anticipated the inequalities to come concerning electric light in their lives. Although at the time of the fair, electric light in homes was still a luxury attainable by only the very wealthy, its ubiquity throughout the Court of Honor must have given people a sense that its place in everyday life was inevitable. But electric lines wouldn't arrive in ordinary urban and suburban households for decades, and in rural homes for decades after that. Black neighborhoods would be among the last lit in cities — long after electricity had come to seem a matter of course in white neighborhoods — and rural blacks would have an even longer wait than rural whites. The longer they waited for electric light, which would continue to grow ever brighter and become ever more a symbol of modernity, the greater the disparity would seem, for electricity did say yes or no with the same voice: the lines ran — or did not run — along the streets and into the homes; electric light suffused entire windows (whereas oil lamps did not). Thus the distinction between those with and those without would come to be as pronounced as the gulf between the Midway Plaisance and the Court of Honor.

At that time, however, when electricity in the home was out of reach for almost everyone, more visitors wandered among the exhibits in the Electricity Building than at any other exposition site, especially in the evening, when it was the brightest place in the White City. After visitors walked past the statue of Benjamin Franklin — "his gaze turned upward toward the lowering clouds, in one hand the kite, and in the other the key of which all the world has read" — they encountered the General Electric exhibit, with its displays of Edison's phonograph and his Kinetoscope, which continually projected a short film of British prime minister William Gladstone addressing the House of Commons. Beyond, visitors could peruse twenty-five hundred specimens of Edison incandescent lamps — "no two of which were alike, being in many colors and in candle power ranging from ½-c.p. to 300-c.p" — and other lamps in different stages of construction, as well as examples of the filaments that Edison carbonized in his quest for incandescent light and examples of his dynamos. At the center of it all stood Edison's Tower of Light, an eighty-two-foot-high column built of thousands of miniature colored lamps that flashed in various designs. It was crowned by a huge incandescent bulb made of cut glass.

The attempt to subdivide electric light may have continued for almost a century and have involved dozens of experimenters and electricians, but Americans would always think of Edison as the sole inventor of the electric light, and he would always hold a particular and sentimental place in the popular imagination, as was clear during the opening ceremonies of the Electricity Building. One observer wrote:

> The Edison tower and the classic pavilion at its base stood revealed in all their cold, chaste beauty of outline. But for a few seconds only; the glare of search-lights focused upon them, causing their dark surface to shine with a dazzling radiance. Then the crystal bulb at the top burst into flame, flashing like a crown of diamonds; and finally the entire column was arrayed in robes of purple light like a pillar of fire . . . and by a thousand voices was shouted the name of him by whom these marvels had been wrought.

Beyond the General Electric exhibit, in displays of both foreign and domestic electric manufacturers, visitors encountered countless things that would have been unimaginable twenty years before: motors, engines, welding equipment, surgery and dentistry instruments. "Close at hand one may study the system of an electric signal company . . . in a neat railway model marked 'dangerous'; he may have a suit of clothes cut by an electric machine, or he may seat himself in an easy chair while his boots are polished by electric brushes. Here also is an electric incubator, with eggs in the process of hatching." They marveled at the exhibit of an electric kitchen, where flameless heat for cooking turned on instantaneously, water poured from a faucet at the turn of a knob, and machines washed clothes and dishes. But electricity not only offered a vision of the future; it also seemed to redefine history. Dioramas depicted past civilizations retrofitted with electricity, such as Egyptians "dipping reels of wire into insulating baths, and bearing to their queen, typical of Chicago, lamps, dynamos, motors, batteries, and other appliances."

The controlled use of power and light was but a part of electricity's allure. The mystery and seeming wildness of it was another, and that was Nikola Tesla's province. The Tesla exhibits in the Westinghouse section of the Electricity Building, inexplicable to almost everyone who saw them, included Tesla's whirling "egg of Columbus" — a copper egg spun on its end within a rotating magnetic field. Lightning crackled between two insulated plates, and various balls and disks spun simultaneously in different places in the room. "When the currents were turned on and the whole animated with motion, it presented an unforgettable spectacle," recalled one witness. "Mr. Tesla had many vacuum bulbs in which small light metal discs were pivotally arranged on jewels and these would spin anywhere in the hall when [an] iron ring was energized."

Tesla also displayed various discharge lamps, offspring of the Geissler tube — the creation of Heinrich Geissler, physicist and maker of scientific instruments in Bonn, Germany, in the mid-nineteenth century. Geissler had taken an evacuated glass cylinder and attached electrodes to both ends, then filled the cylinder with combinations of rarefied gases such as neon and argon. The gases conducted a current from one end of the tube to the other, producing visible colored light in the process. Tesla shaped tubes of light into coils, circles, and squares and spelled out the names of famous electricians, the name of his favorite Serbian poet, and the very word "light."

More intriguing than all his devices was Tesla himself, frail and hollow-cheeked from an exhausting year of ceaseless work. When he arrived at the fair to give a lecture, even the professors gazing at the hodgepodge of equipment he was about to use in demonstrations "lumped off the whole lot under the generic term of 'Tesla's animals.'" Tesla, the announcement for his lecture proclaimed, promised to pass a current of 100,000 volts through his body, "without injury to life, an experiment which seems all the more wonderful when we recall the fact that the currents made use of for executing murderers at Sing Sing, N.Y., have never exceeded 2,000 volts." The promise drew crowds of people clamoring to get into the auditorium, although the demonstration was open only to members of the International Electrical Congress, who were convening at the fair.

Whereas low-frequency current would have meant certain death, Tesla, dressed in a white tie and tails, employed a very high frequency current, which traveled along the surface of his body, not through it. He explained:

> The streams of light which you have observed issuing from my hand are due to a potential of about 200,000 volts, alternating in rather irregular intervals, sometimes like a million times a second. A vibration of the same amplitude, but four times as fast ... would not burn me up. ... Yet a hundredth part of that energy, otherwise directed, would be amply sufficient to kill a person. ... The amount of energy which may thus be passed into the body of a person depends on the frequency and potential of the currents, and by making both of these very great, a vast amount of energy may be passed into the body without causing any discomfort.

Those lucky enough to get a seat were astonished to witness him onstage engulfed in light and perfectly sensible. One reporter of the time wrote: "After such a striking test, which, by the way, no one has displayed a hurried inclination to repeat, Mr. Tesla's body and clothing have continued for some time to emit fine glimmers or halos of splintered light." Engulfed in light is how he is imagined still. Photographs of Edison, whose successes inched forward by dint of ceaseless trial and error, depict him posing with his crew, or perhaps napping on a laboratory table, the background cluttered with bottles and vials and hand tools. The most renowned photographs of Tesla show him alone and somehow saturated with electricity. One, a double exposure, portrays him calmly seated in his spare and cavernous Colorado lab while jagged streaks of light shoot through the air above and around him.

For all the wildness of Tesla's exhibits, his greatest accomplishment at the fair stood in Machinery Hall: twelve perfectly synchronous polyphase dynamos — each about ten feet high and weighing seventy-five tons — which sent current to every corner of the grounds. The hall, Jill Jonnes, notes, was "alive with the deafening mechanical clanking and whirring . . . and unpleasantly redolent of fumes and oil and grease. . . . Great engines in the Westinghouse nave ran even greater generators, which in turn flashed 2,000 volts of AC from each double Tesla machine forth through the subways." But it wasn't the machinery alone that held people's attention. "Popular interest was divided between these machines, the largest of their kind up to the time, and the switchboard," wrote Westinghouse biographer Francis Leupp. That switchboard, made of a thousand square feet of marble and situated in a gallery that was reached by spiral staircases, controlled 250,000 incandescent lights. "What astonished visitors most, perhaps, was to see this elaborate mechanism handled by one man, who was constantly in touch, by telephone or messenger, with every part of the grounds, and responded to requests of all sorts by the mere turning of a switch."

One of the most provoking testimonies to the quality of

the light controlled by the turning of that switch is Winslow Homer's *The Fountains at Night, World's Columbian Exposition*, which he painted while visiting the White City. For centuries, artists had depicted nights in warm, muted colors and worlds disappearing in shadow where a viewer might sense the ongoing fading of light. But in Homer's work, the illumination feels endless. This isn't the light of antiquity: flowing and falling water spans the picture, and the light has turned it entirely luminous — no wake of light in the dark here — against which the statuary and the gondola with its oarsmen and passengers seem all that much darker. Bright white flecks the forelocks, foreheads, and noses of the resolute horses of Frederick MacMonnies's fountain and the upturned face of one of the fairgoers in the gondola that swiftly cuts across the lake. It feels as if the boat will momentarily race past the frame of the painting — it is we who are ephemeral — but the light will never change, or so it seems where the work now hangs, in the midst of other nineteenth-century oil paintings. Surrounded by a nimbus of rich reds, browns, and greens; by paintings of pastures and marshes in pure daylight or slowly disappearing in the dusk; by depictions of fruit and wood and faces kindled by oil lamps and toned down by varnish and time, Homer's *Fountains* — with its blacks and whites and grays, its grave intensity — stands at odds with everything else in the room, as if he has painted the unblinking eye of the late century staring into the future.

When the World's Columbian Exposition went dark after six months, the Laplanders put oceans between themselves and the Dahomey, the belly dancers, and the sword fighters. The buildings, sheathed with "staff" (plaster of Paris mixed with jute fiber and cement), had always been mere spider weaves of struts and supports meant to last only for a summer and fall. Both the mayor of Chicago and the architect of the White City, Daniel Burnham, advocated burning the grounds. "I believe," the mayor said, "if we cannot preserve it . . . I would be in favor of putting a torch to it . . . and let it go up into the bright sky to eternal heaven." Although some of the buildings were destroyed by an accidental fire in 1894, much of the fair was dismantled. "There are 'bits' of the World's Fair at the present time all over the world — in Europe, in Asia, in Africa, in the two Americas, in Australia," reported *Scientific American*. Some of the plaster ornaments were sold as souvenirs; some of the glass went to greenhouses; the salvaged steel was sent to Pittsburgh furnaces. Flagpoles ended up at schools and convents. The statue of Benjamin Franklin found a home at the University of Pennsylvania.

The stuff of the fair may have been dissipated across the globe, but the brilliant, unbounded lights of the Court of Honor would not be forgotten. It seemed that forever afterward, Americans would prize more and more dazzle in their cities, prize electric marquees and electric advertising on an outsize scale made all the more possible by George Westinghouse's next project. Long before the last bits of the White City's plaster were sold off, Westinghouse turned his attention to Niagara Falls, where with the help of Nikola Tesla's dynamos, he would develop the first extensive and practical long-distance power lines.

NIAGARA: LONG-DISTANCE LIGHT

B Y THE TIME Charles Dickens visited Niagara Falls in 1842, it was already thick with visitors. Taverns, viewing towers, stairways, and hotels sprinkled the banks, but their presence couldn't diminish his astonishment:

> I was in a manner stunned, and unable to comprehend the vastness of the scene. It was not until I came on Table Rock, and looked — Great Heaven, on what a fall of bright-green water! — that it came upon me in its full might and majesty. . . . Then, when I felt how near to my Creator I was standing, the first effect, and the enduring one — instant and lasting — of the tremendous spectacle, was Peace. Peace of Mind: Tranquility: Calm recollections of the Dead: Great Thoughts of Eternal Rest and Happiness: nothing of Gloom or Terror. Niagara was at once stamped upon my heart, an Image of Beauty; to remain there, changeless and indelible, until its pulses cease to beat, for ever.

Changeless in Dickens's heart, perhaps, but although the State of New York — interested in maintaining the natural beauty and the appeal of the place for tourists — preserved the area directly around the falls from industrial development, Niagara contained so much exploitative potential that it could not possibly remain changeless in the industrial age. Nineteenth-century magnates believed that its power was lying in wait for them, if only they could arrive at a way to harness the force of the water. In the words of inventor Sir William Siemens, "All the coal raised throughout the world would barely suffice to produce the amount of power that continually runs to waste at this one great fall."

Niagara's 160-foot precipice of dolostone and shale isn't among the highest of cataracts, but with a breadth of more than 3,500 feet, it is second only to southern Africa's Victoria Falls in width. And the lakes that feed into the Niagara River — Superior, Huron, Michigan, and Erie — contain 20 percent of all the fresh water in the world. When Swedish traveler Peter Kalm encountered the river in 1750, almost all of it rushed over the falls and then through a series of gorges before flowing into the fifth Great Lake, Ontario. "The greatest and strongest battoes would here in a moment be turn'd over and over," he wrote. "The water . . . seems almost to outdo an arrow in swiftness. . . . When all this water comes to the very Fall, there it throws itself down perpendicular! It is beyond all belief the suprize when you see this! . . . You cannot see it without being quite terrified."

When Kalm visited Niagara, the rugged, lush country — its vines, flowers, mosses, and pines drenched by mists rising from the falls — held few human traces beyond the cold fires of old encampments and the portage and trading paths of the Iroquois. The river, far too vast and swift to navigate, was mostly an obstacle to the tribes in the region, though they sometimes gathered up fish that perished in the roil at the bottom of the falls — the drop being deadly to all kinds of wildlife caught in the currents. Kalm wrote:

> Several of the *French* gentlemen told me, that when birds come flying into this fog or smoak of the fall, they fall down and perish in the Water; either because their wings are become wet, or that the noise of the fall astonishes them, and they know not where to go in the Dark. And very often great flocks of [swans, geese, ducks, water-hens, teal, and the like] are seen going to destruction in this manner; they swim in the river above the fall, and so are carried down lower and lower . . . till the swiftness of the water becomes so great that 'tis no longer possible for them to rise, but they are driven down the precipice, and perish. . . . They find also several sorts of dead fish, also deer, bears, and other animals which have tried to cross the water above the fall; the larger animals are generally found broken to pieces.

The Europeans and Americans who settled northern New York in the eighteenth century, like the area's native peoples, found the power of Niagara far too great to exploit. As they cleared the woods and planted fields and orchards, they instead dammed small area streams and rivers for their sawmills, gristmills, and carding machines. The one village — a tavern, a blacksmith, and a handful of homes — along the river above the falls, where a narrow canal fed a small sawmill, then a gristmill, burned to the ground during the War of 1812. A new community, eventually called Niagara, established itself on the ruins of the old town, and several small mills there were driven by water channeled through a canal.

More extensive exploitation of the power of Niagara would prove to be an extraordinarily complex undertaking that would take almost a decade of concerted effort, enormous capital, and an investment in untried technologies. It began in 1886, when Thomas Evershed, the divisional engineer of the New

York State canal system, conceived a plan to build a water-wheel power system at the falls upstream from the preserved land. He envisioned a series of lateral canals, which would turn numerous waterwheels in an industrial complex of mills and factories. A two-and-a-half-mile tunnel that ran directly underneath the town of Niagara would return the water to the river just below the falls. But even if Evershed could sell power to several hundred businesses clustered near the falls, it wouldn't carry the cost of the endeavor. To make a profit, he would have to find a way to transmit Niagara power over twenty miles to Buffalo — a city, then, with a population of a quarter of a million people — where it could provide electricity for manufacturing, the trolley system, and public and domestic lighting. At the time, neither alternating current nor direct current could send electricity over more than a few miles.

Evershed had trouble attracting investors to his risky endeavor, and three years later, finding himself strapped for funds and unable to raise money, he turned over the project to Edward Dean Adams, a New York banker. Rather than building a series of lateral canals, Adams envisioned constructing a central station along the falls, from which electricity would be sent to industries in the area and then eventually to Buffalo. Although the plan was just as untried as Evershed's, Adams was a respected financier and managed to attract the interest and investments of some of the richest businessmen of the time, including J. P. Morgan, John Astor, and William Vanderbilt.

In October 1890, Adams began work on the tailrace, which would carry water away from the turbines and would be necessary for any approved design. Everything about the endeavor was massive: "Thirteen hundred workmen were blasting their way, day and night, through the solid rock, 160 feet below the town," notes Niagara historian Pierre Berton. "The horseshoe-shaped tunnel, eighteen feet wide, twenty-one feet high, and seven thousand feet long, would displace 300,000 tons of rock; it would require twenty million bricks to line it and two and half million feet of oak and yellow pine to shore it up." Yet even as construction was proceeding, Adams had no solid idea as to how to transport the power over a long distance. He conducted an international competition of electricians and engineers in an attempt to find a means for transmission. Plans proposing the use of alternating and direct current were submitted, but no feasible proposal came of the challenge.

To efficiently and cost-effectively transmit power over long distances, any system would have to rely on high voltages: the flow of current would increase, but the resistance would remain constant. High voltages—too high to run lights or motors—have to be transformed: that is, they have to be stepped up to higher voltages when leaving the generators and entering the lines, then stepped down to lower voltages before arriving in homes or factories. Whereas direct current could not be transformed (transformers rely on an oscillating magnetic field, and direct current flows only one way), alternating current could. Although a transformer for alternating current had already been developed, alternating current was still virtually untested for long-distance transmission. The only proof of its feasibility lay in an experimental system built in 1891 in Germany, which had transmitted power from Lauffen to Frankfurt—a distance of more than a hundred miles—to run machinery and lighting at an electrical exhibition, and at the Gold King Mine in Telluride, Colorado, where a Tesla polyphase generator transmitted power two miles to run a motor in the crushing mill.

In late October 1893, in part because of the success of alternating current at the World's Columbian Exposition, Adams awarded George Westinghouse a contract to build the first generators at Niagara. And Westinghouse turned to Tesla. Niagara had been in Tesla's mind since he'd seen a steel engraving of the falls as a teenager. He later wrote, "[I] pictured in my imagination a big wheel run by the Falls. I told my uncle that I would go to America and carry out this scheme. Thirty years later I saw my ideas carried out at Niagara and marveled at the unfathomable mystery of the mind."

By 1895 Adams had installed, within a cavernous brick powerhouse designed by Stanford White, called the Cathedral of Power, three 5,000-horsepower Tesla polyphase generators (those powering the White City had been 1,000 horsepower), each weighing eighty-five tons. They endured countless tests and were repeatedly calibrated and reset, and in August of that year "the inlet gates at the canal opened, the river water flooded into one of the penstocks, the turbine whirled, and so did Dynamo No. 2, flashing alternating current off to the Pittsburgh Reduction Plant [a nearby aluminum manufacturer]." Upon the successful transmission of power, Tesla predicted that "the falls and Buffalo will reach out their arms and will join each other and become one great city. United, they will form the greatest city in the world."

The next year, at one second after midnight on November 16, 1896, the switches were pulled at the power station at Niagara, and the current, stepped up through transformers, ran along twenty-six miles of cable and was then stepped down and delivered to the streetcars of Buffalo. "Electrical experts say the time [it took] was incapable of computation," remarked one reporter. "It was the journey of God's own lightning bound over to the employ of man." Within months, Niagara began to supply current that lit Buffalo's streets, homes, businesses, and industries.

Here was power untethered from its source, freed from the lay of the land and the flow of rivers, abstract and seemingly without limit. "Wherever mankind wishes to go," one observer later wrote, "copper wires can go, too." But the technical ability to send energy across long-distance wires also brought with it all kinds of new challenges. Electric companies would need to refine the process by which they delivered power and adapt to the demands of its users — or force users to adapt to their desires. Societies would need to confront the disadvantages experienced by those beyond the reach of electric power, and people everywhere would have to come to terms with living with the inexplicable. "Yoked to the Cataract!" the *Buffalo Enquirer* proclaimed, which also meant yoked ever increasingly to something not even the great inventors understood. "What *is* electricity?" one writer of the time inquired. "That is a question no man can yet fully answer. . . . The men who make the dynamos and the men who operate them know how to produce electricity, but Mr. Edison himself, standing by an Edison dynamo, could only tell you the 'how,' and not the 'why.' Yet, for thousands of years this great power has been in the universe, waiting for nineteenth-century man literally to find it out."

Even Tesla was never able to adequately explain electricity:

Now, I must tell you of a strange experience which bore fruit in my later life. We had a cold [snap] drier than ever observed before. People walking in the snow left a luminous trail. [As I stroked the cat] Macak's back, [it became] a sheet of light and my hand produced a shower of sparks. My father remarked, this is nothing but electricity, the same thing you see on the trees in a storm. My mother seemed alarmed. Stop playing with the cat, she said, he might start a fire. I was thinking abstractly. Is nature a cat? If so, who strokes its back? It can only be God, I concluded. . . . I cannot exaggerate the effect of this marvelous sight on my childish imagination. Day after day I asked myself what is electricity and found no answer. Eighty years have gone by since and I still ask the same question, unable to answer it.

Here was light taken on faith and perhaps replacing faith. Man of letters and historian Henry Adams understood the true significance of the dynamo: "To Adams the dynamo became a symbol of infinity. As he grew accustomed to the great gallery of machines, he began to feel the forty-foot dynamos as a moral force, much as the early Christians felt the Cross. The planet itself seemed less impressive, in its old-fashioned, deliberate, annual or daily revolution, than this huge wheel, revolving within arm's length at some vertiginous speed, and barely murmuring." And why wouldn't it? One moment our world was dark, and the next brilliant. That almost no one understood how this was accomplished, and that this light, unrelated to the eons of tallow and coal; this light, requiring nothing of us — no fussing over a flame or wick, no worry over the quality of the oil; this light, with its own particular trajectory tied to the precision of the industrial age — timed and tuned, and pitched and keyed, all rhythm and exactitude; this light, conjured by wizards — as both Edison and Tesla with their varying temperaments were called; this light, with its constancy and brilliance, was nothing if not the evidence of things unseen.

What it did require, of course, was that we go forward on trust. What culminated at Niagara was only the beginning: the electric grid would come to be considered the greatest technical accomplishment of the twentieth century. New wizards would detach us even more from the things of this earth, and we would need to trust also that our data, words, and life's work would not in an instant disappear from before our eyes. H. G. Wells understood that something fundamental had shifted as he stood looking at the falls in 1906. Not only had the spiritual been fused to the industrial, but it also seemed that some glory had been taken away from Nature herself. He wrote:

> The dynamos and turbines of the Niagara Falls Power Company, for example, impressed me far more profoundly than the Cave of the Winds; are indeed, to my mind, greater and more beautiful than that accidental eddying of air beside a downpour. They are will made visible, thought translated into easy and commanding things. They are clean, noiseless, and starkly powerful. All the clatter and tumult of the early age of machinery is past and gone here; there is no smoke, no coal grit, no dirt at all. The wheel-pit . . . has an almost cloistered quiet about its softly humming turbines. . . . The dazzling clean switch board, with its little handles and levers, is the seat of empire over more power than the strength of a million disciplined, unquestioning men.

PART III

So if we moderns were to enter into an interior of the past, we would very soon feel uncomfortable. However beautiful it might be — and it was often wonderfully so — what for them exceeded sufficiency would not be enough for us.

— FERNAND BRAUDEL,
Capitalism and Material Life, 1400–1800

New Century, Last Flame

In our households we talk of dynamos, motors, trolleys, electric lamps, telephones, and batteries, quite as freely as we do of bread, butter, butcher's meats, milk, ice, coal and carpets.

— EDWIN J. HOUSTON,
Electricity in Every-Day Life (1905)

THE TALK IN HOUSEHOLDS may have included all manner of things electric, but when H. G. Wells stood at Niagara in 1906, electricity was still confined to dense urban areas, and within those areas it was available almost solely to businesses, manufacturers, and wealthy homeowners. Even so, city dwellers had grown accustomed to electric light in public, and although most still used non-incandescent lamps at home, almost all light was measurably cheaper and more efficient than it had been in the past. Gas, for instance, sold for $2.50 per thousand cubic feet in 1865. By the end of the nineteenth century, it cost about $1.50 per thousand cubic feet. And kerosene, which sold for about 55 cents a gallon in 1865, dropped to as low as 13 cents a gallon by 1895. Only commercially produced tallow candles — rarely used in the late nineteenth century — had become more dear. Whereas in the early nineteenth century, they could be purchased for 20 cents a pound, in 1875 they cost 25 cents a pound.

Consequently, most American households at the turn of the twentieth century were much brighter than those of the past. In 1800, in the United States, $20 a year would light a house for three hours in the evening with a luminosity equivalent to 5 candles, or 5,500 candle hours per year — and many householders would have considered that much light an extravagance. By mid-century, $20 would purchase 8,700 candle hours per year; in 1890, 73,000 candle hours. By 1900, for $20 a year, on average people lit their homes (exclusive of electricity) for five hours a night with a luminosity equivalent to 154 candles, or 280,000 candle hours. That miners once worked by the phosphorescence of putrescent fish and lacemakers produced intricate designs by the light of a flame magnified through water must have seemed incomprehensible to them.

It's worth remembering that this rapid increase in the ease and brilliance of light was limited to industrialized countries. Millions throughout the world knew nothing of electricity, gaslight, or even kerosene. Their illumination, both in substance and meaning, had changed little since ancient times. Perhaps nowhere did traditional lighting hold more meaning than in the high latitudes where the Eskimo, Inuit, and other northern peoples — their villages dispersed across the snow and ice, and they themselves outnumbered by the animals — lived for months at a time with scarce daylight. Richard Nelson describes the Koyukon people of the Alaskan interior:

> Houses were lit by burning bear grease in a shallow bowl with a wick, or by burning long wands of split wood, one after another. Bear grease was scarce, and the hand-held wands were

inconvenient, so in midwinter the dwellings were often dark after twilight faded. Faced with long wakeful hours in the blackness, people crawled into their warm beds and listened to the recounting of stories. . . . The narratives were reserved for late fall and the first half of winter because they were tabooed after the days began lengthening. Not surprisingly, the teller finished each story by commenting that he or she had shortened the winter: "I thought that winter had just begun, but now I have chewed off part of it."

For those in the northernmost coastal villages of Greenland, Canada, and Alaska — where in the heart of winter, the only natural light comes from the stars, the moon, and the aurora borealis and the only source of fresh water is locked in snow and ice — stone lamps were utterly essential for survival: among the Inuit of Greenland, "the constellation of the Great Bear is called . . . *pisildlat*, lamp foot or stool upon which the lamp is placed."

Above the tree line, where only occasional driftwood might be available for fires, people relied almost entirely on seal oil, a more efficient fuel than reindeer fat or the fat of other land animals. Women carefully gathered every last bit of it from the carcass, scraping the skin with an ivory scoop, and they saved any oil that might drip from the lip of their lamps, which were carved of soapstone. The exact size and shape of the lamps varied from village to village, but most were elliptical — a foot or two long — with a thick edge. A wick of dried moss, catkins, or peat — rubbed between the palms with a bit of fat — would be laid in a thin line along the edge. The lamp could be tipped to feed more oil to the wick. Sometimes a slab of seal blubber hung over the bowl and fed more fat to the lamp as it melted.

If more than one family shared a snow shelter, as often happened, each possessed its own lamp, which kept family members warm and cooked their food. Its heat also dried their clothes and boots and was used to tan hides. Steam rising off the cooking pots helped the people to bend straps of wood and pieces of bone, from which they fashioned snowshoes and boxes. Most essentially, it gave them water to drink. Humans can't eat snow — it isn't high enough in water content to prevent dehydration before it lowers the core body temperature to fatal levels. So those living in the farthest north had to melt snow for their drinking water, either directly over the flame or near it, where a chunk of snow or ice might lay on an inclined slab, its meltwater slowly running into a container.

As the lamp burned, it warmed the cold air coming through the entryway of an ice house; the heat rose and escaped through a vent in the ceiling. The walls continually thawed and froze, thawed and froze. When people placed animal skins over the interior walls to keep them from dripping, the lamp might throw enough heat that family members could sit shirtless in the house. In small, low ice houses, the lamp might smoke as the family slept, and they would wake covered with soot, suffering from headaches, and starved for oxygen. In the late 1960s, when scientists at the Walter Reed Army Institute of Research examined the mummified remains of an Aleut (Aleutians also used seal oil lamps), they found the lungs to be coated with a thick, black substance. One of the scientists said, "Had he smoked I would have called him a three pack a day man."

However smoky, the lamp meant so much to families that in lean times, so as to have enough fuel for the fire, they were willing to go hungry. The fire was almost always kept alive, most often carefully guarded and tended by the woman of the household. She spent much of her day alongside it, cooking, preparing hides and skins, sewing winter clothing, and drying clothes. The flame, a few inches high, was difficult to keep clear and smokeless. In the late 1800s, anthropologist Walter Hough noted: "Lamp trimming only reaches perfection in the old women of the tribe, who can prepare a lamp so that it will give a good, steady flame for several hours, while usually half an hour is the best that can be expected. In an Eskimo tradition a woman takes down some eagle's feathers from a nail in the wall and stirs up the smoking lamp, so as to make it burn brightly." Elsewhere he wrote, "The Eskimo have no phrase expressing a greater degree of misery than 'a woman without a lamp.' After the death of a woman her lamp is placed upon her grave."

For those living in the early-twentieth-century cities of Europe and America, the regard the inhabitants of the circumpolar regions had for their soapstone lamps might have been as hard to fathom as the meagerness of the flame. In these cities, any open flame, however bright, had become easy to disparage and at best carried a hint of nostalgia. All the improvements — the twisted rag becoming a plaited wick, Argand's steadying of the flame, the clarity and brilliance of kerosene and gas light — would soon be no more than history, and the lamp's mysteries would be memory's mysteries, as essayist and critic Walter Benjamin knew. In the 1930s, Benjamin remembered the lamp of his childhood, a lamp that,

> unlike our lighting systems, with their cables, cords, and electrical contacts, you could carry . . . through the entire apartment, accompanied always by the clatter of the tube in its casing and the glass globe on its metal ring — a clinking that is part of the dark music of the surf which slumbers in the laborious toil of the century. . . . Now the nineteenth century is empty. It

lies there like a large, dead, cold seashell. I pick it up and hold it up to my ear. What do I hear? . . . The rattling noise of the anthracite that is emptied from the coal scuttle into the furnace; . . . the clatter of the tube in its casing, the clink of the glass globe on its metal ring when the lamp is carried from one room to another.

Soon most would forget how to light a lamp and how to husband the flame. They would become a little afraid of it, and it was its obviousness that seemed dangerous: its smell, substance, and centuries of meaning. How could a simple flame hold a plea against electricity? Aggressive, unhindered electricity: "Let's Kill the Moonlight!" proclaimed the Italian futurist poet Filippo Marinetti, for he saw the natural world as irrelevant, canceled by the speed and brilliance of the modern. Giacomo Balla, in his 1909 oil painting *Arc Lamp*, seems to have done just that. Artificial light dominates everything; even the iron base of the streetlamp has relinquished its solidity. It's just a ghost, clouded by the sizzle of energy — circular, radiant, pulsing, full of hot color — flaring out from the arc. The light's sharp power and verve strike at soft, billowing, susceptible night. There is hardly any room left for the dark, which is trying to hold its own in small quarters at the corners of the painting. That's more than can be said for the pale crescent moon, helplessly obscured in the background — illuminated, but not radiant, captured, as it is, by human light.

GLEAMING THINGS

T HIS MUCH HAS ALWAYS been true: electricity can't be stored. It must be generated as needed and consumed within moments of its generation. The supply must continually adjust itself to fluctuating demand, and a power plant must have sufficient capacity to meet all its customers' needs at any given moment of the day. Maintaining this balance was especially fraught during the first tenuous decades of electrical expansion. Edward Hungerford, writing about the gas and electric plants of New York City in 1910, described how the smallest change in the skies could create a sudden spike in usage:

> In days of old, watchers were stationed upon the high house-tops of mediaeval cities, to give warning of the coming of an unexpected foe. In these days there are watchers upon the high housetops of the modern city. They go there whenever the barometer begins to spell uncertainty. With powerful glasses they skim the distant corners of the horizon. A distant black cloud — a seemingly harmless thing in the far-away sky, but a thing of magnificent potentialities close at hand — is seen. Its approach is closely noted. . . . The watcher of the skies gives quick warn-

ing over the telephone. The drone of lazy midday ceases instantly. [In the power house] men come out of their drowsy cat-naps. They rush to their positions, fresh fuel goes upon a hundred banked boilers[, and] . . . the 'chief operator,' who is king of the situation, orders additional engines and generators into service. . . . When the black clouds finally rest above the town and the myriad hands are reaching for desklights, the strain has been already met. The light . . . burns as steadily and as brightly as it burned five minutes before, when less than one-fifth the quantity was the demand.

In Hungerford's time, it was also true that electric plants created black clouds of their own, for not all energy generation could be as clean as Niagara. Power plants in places far from any viable waterpower source often relied on coal-fired furnaces to heat water, which produced the steam that commonly rotated the turbines of the generators. And the predominance of alternating current meant that in a city like New York, the hundreds of small local plants that once pocked the city were now consolidated into a few huge generating stations. By 1910 the New York Edison Company plant at Thirty-eighth Street and First Avenue, which replaced four hundred small electric power plants in Manhattan, took up two city blocks and furnished almost 90 percent of the electricity for Manhattan and the Bronx. It ran 152 boilers, which consumed more than half a million tons of coal in the course of one year. The grime and soot the plant produced was a constant source of irritation to neighboring homes and businesses, not only noxious to breathe but also damaging to furniture and draperies. The company, repeatedly fined by the health department for coal smoke violations and cinder nuisances, had watchers of its own. The *New York Times* reported that during an ongoing investigation, "when it was found Health Department men were trying to photograph the smoke stacks, 'scouts' were put on the company's roof who ordered the feeding of coal stopped whenever photographers appeared."

Whatever the fuel source, electric companies have always sought to cultivate a consistent demand for power, since a plant is most efficient and profitable when its output is constant. In the early twentieth century, they courted industrial and commercial customers, who not only used large amounts of electricity at predictable times but also were usually located in concentrated areas, which meant there'd be minimal investment in lines. They particularly sought customers whose demands might complement the municipal drain on electricity from trolleys and street lighting, both of which used a good deal of power early in the morning and later in the day.

Electric companies — still called "light companies" in the early decades of the twentieth century — were private corporations, and since access to electricity was not yet considered the right of every citizen, they felt no obligation to deliver power to individual homes. Electric light in homes, they believed, would exacerbate strains on their systems, since people would turn on their lamps during the peak-demand hours of dusk. It hadn't yet occurred to them to promote the sale of washers, dryers, vacuum cleaners, and irons, which would have increased daytime electric use in homes. At least in the early years of the century, they had little faith that householders would be interested in such things. So by 1912, more than three decades after Edison's Menlo Park demonstrations, only 16 percent of American homes were connected to central station power, and most of those were in wealthy and upper-middle-class districts.

Even in homes wired for electricity, those who wanted to use electric appliances faced a host of obstacles. Household wiring was unregulated and rudimentary, sufficient for little more than lighting alone. The styles and types of plugs varied from manufacturer to manufacturer, and people could plug in smaller appliances only if they had the correct outlets for them. If a family purchased a stove, which required insulated wires, or a refrigerator, which ran on higher-than-normal wattage, they usually had to upgrade the wiring in their home. As late as 1926, one commentator observed: "Electrical articles are the only ones which cannot be taken home and put to use by the purchaser, when, where, and as he pleases!"

The quality and design of many early appliances was poor as well. One man, recalling his mother's first iron, noted: "It was a Dover iron. And even though it had a plain, unplated iron soleplate and a nickel-plated shell, we thought it looked pretty swell. . . . The new iron did a wonderful job. But the attached cord, which ran directly inside the shell to the terminals, kept burning off because of the heat at that point." There were no safety standards and few guarantees. When appliances broke down, as they often did, there was no system of service for repair. What was a householder left with? Often no more than a "so-called instruction booklet which never in eight years has helped us in a single emergency. . . . Does the motor stop, the engine refuse to start, is there a mysterious 'spark,' 'smoke,' unexplained 'knock' — we can pore through the booklet in vain for help."

Even so, the marvel and mystery of it all was very much alive, however unrealistic and unattainable. Manufacturers continued to demonstrate electricity's promise at world exhibitions and in model electric homes outfitted with clothes washers and dryers, dishwashers, stoves, and refrigerators. Books such as *Electricity in Every-Day Life* and *Electric Cooking, Heating, Cleaning, Etc., Being a Manual of Electricity in the Service of the Home* gave readers a brief history of electricity and explained how it would inevitably revolutionize their lives. One author exclaimed, "Fancy cooking cutlets and frying pancakes with captured lighting!" Such books promoted electricity not only as a timesaver for women but also as a replacement for domestic help, which had become scarce as workers increasingly chose more lucrative and independent work in factories over domestic employment. One advocate of appliances proclaimed: "There is no household operation capable of being mechanically performed, of which, through the motor, electricity cannot become the drudge and willing slave."

Magazine articles declared that the electric life would bring unimaginable ease. In 1904 *Scientific American* published "Electricity in the Household," which described an electric iron, griddle, toaster, and cereal boiler, along with a chafing dish, about which the author claimed: "A traveler will find this stove particularly useful. It can be carried in the overcoat pocket." He also described a sewing machine, the speed of which "can be very delicately regulated. . . . The operator can assume any easy, comfortable position as the only duty required is to steer the cloth under the needle." In the accompanying photograph, a woman dressed, it seems, for a social occasion, is half turned away from her work. Her legs are crossed casually to the side, and she's guiding material toward the needle with her left hand while her other is free and draped over the chair back. She could be chatting with a friend. The author asserts, "Even an invalid can safely operate a machine thus driven."

In these early decades of the twentieth century, electric light bulbs were sold as both a brilliant mystery and a mystery attached to the past. The earliest print ads for them had been straightforward, simply stating their wattage and size. They would often be accompanied by a line drawing of the bulb, base, and filament. But particularly after the development of the brighter, more efficient, and longer-lasting tungsten filament in 1911, the ads became more elaborate. General Electric, still by far the largest manufacturer of light bulbs and lamps, launched a new trademark: Mazda, named after Ahura Mazda, the Persian god of light. Some of the ads for Mazda bulbs featured a reclining woman, draped in flowing robes, who held a light bulb aloft in her outstretched hand and gazed at the lifted brilliance. The bulb itself glowed without any connection to wires and sockets. Not even the filament was obvious, as if to suggest that the new was not so different from the old after all, for nothing in the ads hinted at the way light was now tethered to a growing industrial grid.

Electric lines eventually made their way into middle-class urban and suburban neighborhoods, spurred by Samuel Insull's adoption of the demand meter in Chicago. The meter encouraged use because it allowed power companies to charge lower rates to customers who consumed more than the minimum amount of electricity. Insull, as president of Commonwealth Edison in Chicago, had foreseen increased domestic demand for electricity and actively sought out suburban customers, offering cheap wiring for their homes. Historian Harold Platt notes that Insull "went after every kind of customer from the biggest to the smallest. Maybe the smallest was the household and the housewife. In one famous campaign, he brought in 10,000 GE irons and gave them away free, so to speak, to anyone who would sign up for service."

When electricity finally arrived at their doors, families usually bought smaller appliances first, though not entirely because they were less expensive and easier to bring into their homes than larger ones. A refrigerator wasn't all that essential in a time when corner stores flourished — women shopped almost daily — and milkmen came to the door. As well, the advent of refrigerators spurred icebox manufacturers to improve their goods, and icemen stepped up their home delivery service. As for stoves, gas had already revolutionized cooking for city women. They didn't have to load fuel or tend a flame, and each individual burner operated at the turn of a switch, so they could use one burner at a time rather than heat up the entire stove for a can of soup or a tin of beans. Tin cans had come into their own by then, though there were no standards there either. As Christine Frederick observed, "A tin can is literally a dark, sealed mystery until it is opened."

Women knew what they wanted, and as Insull had foreseen, most purchased an electric iron first. The ads for them always showed a contented, well-dressed housewife effortlessly running an iron over her family's clothes. This was a stark contrast to the old chore, for there was no greater symbol of household drudgery than the "sad-iron" — "sad" in its archaic sense, meaning "heavy" or "dense." Traditional irons, made of cast metal, usually weighed four or five pounds, though some weighed as much as ten. The heavier the iron — and the more a woman pressed down on it — the more efficiently it worked. On ironing day, a woman would heat four, five, or six irons on her gas or wood stove. Before using one of the hot irons, she'd wipe the bottom clean, rub it with beeswax, and try it on an old piece of cloth to make sure it wasn't so hot that it would scorch the cloth. Then she'd press it onto a Sunday shirt, all the while taking care not to transfer any soot to the clean shirt and not to burn herself or the cloth. Once off the heat, the iron would cool down quickly, and in no time at all she'd have to return it to the stove and replace it with a hot one, which she would clean, wax, test . . . Given the mountains of wrinkled cotton clothes and linens to be ironed, the job would take all day. And all the while, the woman would be standing next to the hot stove, even in high summer. One electric iron replaced every sad-iron in the household, and not only did it save time, but it was also far cleaner and more predictable, since the iron kept a constant temperature.

After irons, women most frequently purchased vacuum cleaners. Electricity was sometimes called "white coal," part of its allure being that all the attendant work and grime of production existed somewhere out of sight, so that people could believe the claim that "electricity, the unseen and the unknown, is absolutely clean." While electricity didn't produce the household smoke and residue of gaslight or kerosene lamps, what dirt there was now lay exposed by the increased candlepower of the tungsten filament, and dirt seen was dirt that had to be dealt with.

> Woman has been a dirt eraser for so many ages with no relief in sight and no hope of anything better than beginning again at the moment of finishing. . . . [The vacuum cleaner] has a gigantic value in lifting the woman from her long and seemingly doomed relationship with dirt in the wrong place. . . . The machine that removes it, sucks it right out of the house altogether . . . and is used in the average home about two hours a week. The old broom had at least a half day record. About the same intelligence is needed in the operation of each, although the cleaner requires far more thought and care to keep it in fitness and can be as successfully handled in a dinner or calling costume as with apron and cap.

It was a boon to all but the broom makers, who were taken to task by one advocate of sweeping as he made a desperate pitch for tradition: "They have let go, unchallenged, that sweeping is drudgery until the present generation thinks and talks of sweeping as menial labor, unpleasant and to be performed with reluctance. What a misconception! The medical profession in numerous instances advises women to take up housework, especially sweeping, to offset their ills. Sweeping is exercise of a highly beneficial nature." He was a voice crying in the new wilderness.

In this new wilderness, nothing was more complicated than time. But time — though no less an obsession than cleanliness — being abstract and malleable, couldn't be confronted in a straightforward way. Within more affluent homes in the first decades of the twentieth century, women were often thought to have too much time on their hands. *Ladies' Home Journal* declared: "As a matter of fact, what a certain type of woman needs today more than anything else is some task that 'would tie her down.' Our whole social fabric would be the better for it. Too many women are dangerously idle." But these same women felt pressure to make the most of time. The domestic science movement had taken hold, and its proponents advocated efficiency in household chores, the same way Frederick Taylor, writing in 1911, advocated it for factories: "We can see and feel the waste of material things. Awkward, inefficient or ill-directed movements . . . leave nothing visible or tangible behind them."

Electric appliances could help women be more efficient with housework and brought with them a dream of liberation. But advocates of domestic science believed that efficient work in and of itself was a kind of liberation: "The cry of the home honored woman to be released from the dish pan, the tub and the kitchen range is answered. It is now a matter of how far she will go on the new road and what amount of culture she can and will take on in the performance of the common task.

From a musical standpoint she can move as far as time, tune and rhythm can be made to play upon her daily routine. Artistically we find every effort being brought to bear upon the home to give it the atmosphere it deserves."

Yes, electric appliances saved time. Washing clothes in the age-old way had taken an entire day, traditionally "Blue Monday." With an electric washer, a woman could clean clothes at odd times throughout the week — a load here, a load there — in between other tasks. But for some women, the arrival of electricity ushered in more work than before. The availability of electric appliances put more pressure on women who had relied on domestic help or services to accomplish these tasks themselves. And although the labor of washing had disappeared, so had the community of it. Women who'd previously washed and hung their clothes in the backyard could gossip with hired help or neighbors as they did so. Electric washers and dryers confined them, often alone, to the house. The new efficiency also created new expectations. *Ladies' Home Journal* observed: "Because we housewives of today have the tools to reach it, we dig every day after the dust that grandmother left to a spring cataclysm. If few of us have nine children for a weekly bath, we have two or three for a daily immersion. If our consciences don't prick us over vacant pie shelves or empty cookie jars, they do over meals in which a vitamin may be omitted or a calorie lacking."

Electric light was now but one of many things that made life easier and also seemed to define what it meant to be modern. These things were inextricably linked to the imagination reaching for the future — much like F. Scott Fitzgerald's Gatsby, in the unquiet darkness, stretched toward the single green beacon in the bay who, as young Jimmy Gatz, sought to remake himself: "Rise from bed: 6.00 A.M.; Dumbbell exercise and wall-scaling: 6:15–6:30; Study electricity etc.: 7:15–8:15."

Electric light, however, also brought its own particular changes into homes. Although gaslight had fixed the flame at specific points throughout each room, mantle gas lamps, like kerosene lamps, still provided a living warmth around which people could gather: "When the gas was turned on in the evening, the whole room was bathed in a soft yellow light," remembered one Englishwoman. "Round Aunt Ada's gas mantle was a gas shade made of long crystal glass drops that caught the light and danced like a thousand tiny stars." When gaslight and kerosene lamps disappeared, so did the last vestige of a central fire in the home. Electric light was everywhere, yet concentrated nowhere; everyone sat in the halo of his or her own lamp. The flameless light also brought with it myriad possibilities never before imaginable, since it could be placed where an open flame could not. For instance, one guide to electricity in the home suggested: "In the parlor an illuminated painted vase, lighted from within, may vie in attractiveness with the pictures on the walls, whose colors are almost as readily appreciated by incandescent as by day light, while opalescent globes of varied shade tone the brightness everywhere into subdued harmony. . . . On the veranda the lamps shine heedless of the wind. A very pretty effect can also be produced in conservatories, by suspended lamps of different colors half hidden in the foliage."

Yet electricity did provide a new hearth: the radio, which was also among the most popular electric appliances. The family gathered around voices that broke the membrane between home and the world — voices coming from everywhere, bringing them music, news, weather, farm reports, and preachers. "When they say 'The Radio' they don't mean a cabinet, an electrical phenomenon, or a man in a studio," writer E. B. White said of his community; "they refer to a pervading and somewhat godlike presence which has come into their lives and homes. It is a mighty attractive idol. After all, the church merely holds out the remote promise of salvation: the radio tells you if it's going to rain tomorrow."

By 1920 electric service reached 35 percent of urban and suburban homes. The advent of electric trolleys and the automobile had spurred the move of many middle-class families from the cities to newly built neighborhoods on the outskirts, which had electric service included. Meanwhile, many poor city neighborhoods, home to immigrants from southern and eastern Europe and rural people who'd moved to the city, remained relentlessly in the dark. They had about as much expectation that electric light would come to them soon as an Aleutian Islander did. The social surveys of the time — such as those from Pittsburgh, Pennsylvania, and Lawrence, Massachusetts — which took stock of the deteriorating conditions in crowded city neighborhoods, investigated the lack of natural light, the poor sewage and water systems, and the questionable cleanliness of the milk supply. The surveys didn't mention the dearth of electricity, for not even social scientists yet imagined that access to electricity might be the right of every citizen.

For many immigrant and black city dwellers, the electric life could be quite proximate: poor neighborhoods could exist in the midst of some of the wealthiest sections of a city. In Washington, D.C., they were hidden in plain sight:

> Walk around the outside of this block and you will see nothing peculiar about it. There are two imposing apartment buildings, the former residence of a senator, a handsome club house,

several stylish boarding establishments and a number of three and four story, wholesome private houses. Your attention would have to be directed specifically to the four narrow wagon ways which run inward irregularly from the four sides of the square. A visitor from another city would take these to be passageways merely for the removal of refuse from back yards. But walk a hundred feet down one of these obscure byways and you find yourself on the borders of a new and strange community . . . little wooden or brick houses whose rear doors point toward the rear entrances and separate yards of palatial residences.

David Hajdu, the biographer of jazz composer and pianist Billy Strayhorn, describes the Homewood section of Pittsburgh where Strayhorn grew up: "The whites generally occupied the residences on the main streets — good-sized and well-equipped two-story row houses — and the black families those in the alleys behind them — low-hanging, unpainted shelters with no electricity."

Not only were such neighborhoods dimly lit, but the work many of the women did to make ends meet, such as taking in laundry, was ancient. Clotheslines and washtubs filled the yards. Charles Weller, who documented alley life in the capital in the early twentieth century, described a woman "ironing beside a smoking lamp without a chimney in the front room . . . laying the white, clean-smelling garments into covered baskets for delivery." When Weller approached another woman, he observed that "the perspiring woman was too busy at her wash-tub to waste any time in conversation. She was indignant[.] 'Yes,' she said, 'you folks makes us pay so much rent that we have to scrub our fingers off doing your washing and your scrubbing to earn the money; and we're glad if we can get enough extra to have ash cake and smoked herring for our little ones to eat.'"

Actress and blues singer Ethel Waters, raised in a red-light district of Philadelphia, remembered that "each day was a scuffle, a racking struggle to keep alive. When people are in that situation the problems of a child must seem very unimportant. All that counts is eating and keeping a roof over your head. . . . None of us felt we were underprivileged or victims of society. The families we knew were doing no better than we were, so the daily struggle seemed universal." Still for her, the idea of light, abundant light — be it flame or electric — and its beauty in the night, meant something beyond articulation, no less so than for Fitzgerald's Gatsby. Those with an abundance of it seemed to be living the good life. According to Waters, "The prettiest sight in that whole neighborhood came at dusk when the lights were turned on in the sporting houses. I'd stand on the street and look in with awe at the rich, highly polished furniture and the beautiful women sitting at the windows wearing low-cut evening gowns or kimonos."

12

ALONE IN THE DARK

They are pronounced overhauls . . . the swift, simple, and inevitably supine gestures of dressing and of undressing, which, as is less true of any other garment, are those of harnessing and of unharnessing the shoulders of a tired and hard-used animal.

They are round as stovepipes in the legs, (though some wives, told to, crease them).

In the strapping across the kidneys they again resemble work harness, and in their crossed straps and tin buttons.

—JAMES AGEE,
Let Us Now Praise Famous Men

WITH THEIR LACK of electricity, people living in densely packed poor city neighborhoods could claim close kinship to rural Americans, who, in the first three decades of the twentieth century, also had little expectation that electricity would soon come to them. Electrification of the countryside was an expensive, labor-intensive proposition. Rural lines had to be sturdier than urban ones to withstand the open miles, the winds, the ice and sleet. They could be difficult to string because the lay of the land and the kinds of soil — clay, sandy, stony — varied widely along the routes. Trees had to be trimmed away from the lines. And, since rural lines guaranteed at most only one, two, or three customers per mile — and cautious, parsimonious farmers at that — in contrast to the dozens of customers per mile in cities, electric companies reasoned that rural electrification wouldn't be worth pursuing until all other markets had been fully developed and exploited. If then.

It's not that electricity on the farm hadn't been imagined. During the last decades of the nineteenth century, in Europe especially, scientists had experimented with electric plows, harrows, threshers, pumps, and milking machines. They had constructed electric fences, smudge pots, sheep shears, and prods for balky horses. Electricity, they imagined, would keep down frost, fertilize the soil, milk cows, and destroy weeds. Electric lights would extend harvest days, increase germination, incubate eggs, encourage hens to lay in winter, and keep chicks warm in spring. Electrified rain would spur growth. A farmer one day, it was said, "will be a highly skilled electrician, who from a central switchboard at his farm will direct the germination and growth of cabbages, carrots, potatoes and other crops." And an electrified rural world must have seemed all the more probable after Tesla's motors and transformers created the first long-distance power lines between Niagara and Buffalo. As early as 1895, an article in the *Country Gentleman* predicted that "not improbably the barefoot boy now driving home the cows, when arrived at man's estate, [will] be gaily turning the sod on a Montgomery County farm by a three-share electric plow, with all the power of Niagara Falls behind him."

Yet to turn to the American countryside in 1920 was to part ways with progress. At the time, there were about 6.5 million farms in the United States. Fewer than 100,000 of them were connected to central stations, and most of those were located in small northeastern states near cities, or on the West Coast, where irrigation spurred the development of electric lines. And farmers who were connected to high lines could pay twice as much as city dwellers for service. The lack of electricity only exacerbated the diminishing power of the countryside. Young men and women had been steadily leaving the farms during the past century; by 1920 the depopulation of the countryside reached a point where, for the first time in history, the number of those living on farms and in towns with a population of less than 2,500 was smaller than the urban and suburban population of the United States, which stood at 54 percent. Those actually living on farms (as distinct from those living in small towns) accounted for less than a third of the U.S. population, which meant less money was directed to rural areas for education and health and social services.

The situation would only worsen during the next decade. The high demand for food during World War I had given farmers the incentive to increase cultivated acreage and intensify production. When demand fell off after the war, the markets collapsed, and the price farmers could get for their crops plummeted. With mortgages and loan payments to make, farmers were reluctant to reduce their production, and the continued overproduction of crops only ensured that prices would remain low. Many rural areas fell into a depression a full decade before the stock market crash of 1929.

Invisible beyond the glare, the unending, backbreaking work of farm life continued unabated: "There was no quittin' time and no startin' time — it was all the time." Less than 3 percent of farmers owned tractors; most continued to work their fields with horses, which meant they were still devoting a share of their land to raising feed for draft animals — five acres of oats and hay for each horse. Without electricity, farmers had to haul water for their livestock by hand, and they had to milk their cows by hand, sometimes in the dark — an open flame presented too high a risk in the barn. "You could milk a cow in the dark, but there were a lot of things around a barn you couldn't do in the dark. . . . And that was terrible, to work around a barn with the explodable things — the hay and dust and so forth — with a lantern," recalled one proponent of rural electrification. A farmer from Texas commented, "Winter mornings it would be so dark . . . you'd think you were in a box with the lid shut."

The bottlers required that milk be kept at 50 degrees before pickup. If it wasn't, they rejected it, saying it was good only for pigs. Without refrigeration, farmers had to haul their milk to a stream or a well to keep it cool, or they packed it on ice. New England farmers had cooler weather in summer, and they could cut ice in winter and store it in sawdust, but Southern farmers had to buy ice, which was expensive, and it melted quickly in the extreme heat, even when buried in sawdust.

The lack of central station electric power didn't affect all farms equally. The more prosperous and progressive-minded farmers modernized as they could, independent of the grid. Some generated power with the help of steam engines, windmills, and waterwheels, and in 1912, with the advent of Delco electric plants, which were gasoline-powered generators, more farmers found a little ease. Though expensive to operate, the Delco plants lit the barn for a few hours or pumped water and ran machinery. Almost always, farmers who had them reserved their use for farm work; the household itself remained unchanged. By the time half the residents of cities and large towns were connected to electricity, nearly all rural families still saw by the light of kerosene lamps.

Within the farmhouse, electricity would have made an even greater difference to daily life than in towns and cities, where homes usually had been connected to municipal gas, water, and sewage systems before they were electrified. City wives could take advantage of servants, laundries, bakeries, stores, and butcher shops. For farmwomen, hauling water — a four-gallon bucket weighed more than thirty-two pounds — was one of the more demanding tasks. "I would have to get it . . . more than once a day, more than twice; oh, I don't know how many times. I needed water to wash my floors, water to wash my clothes, water to cook. . . . It was hard work. I was always packing water," commented a Texas farm wife. Another said, "You see how round shouldered I am? Well, that's from hauling the water."

Besides cleaning the house and cooking meals, farmwomen canned fruits and vegetables, which meant hauling wood or coal for the fire and standing over a hot stove almost daily in high summer. When the peaches were ripe, the corn was also ripe, and the beans and tomatoes, and they rotted quickly in the heat. But cooking and harvesting and canning were the least of it. "I have always lived on a farm except the first five years of my marriage, and I think I might almost as soon have been in jail, because the work is so hard and is never done. The hardest is the washing," remarked one woman. Doing laundry not only required hauling and heating water. Farmwomen soaked and scrubbed their entire family's clothes on a washboard in a zinc tub. "I got up many a time at three o'clock in the morning, when I had the family, to wash clothes," one recalled. They heated more water for rinsing, and wrung out all the clothes by hand or put them through a mangle, or wringer, before hanging them to dry. "By the time you got done washing your back was broke," one woman remembered. "I'll tell you — of the things of my life that I will never forget, I will never forget how much my back hurt on washdays." Women then spent another day pressing the family's clothes with their cast-iron sad-irons. Once again, the stove would be heated and enough wood hauled to keep it going all day.

As for light, farmwomen still had to polish the globes of their kerosene lamps once or twice a week and deal with the smoke and soot the flame created. Over time, the brightness of oil lamps had increased — the Aladdin lamp, with its delicate mantle, was advertised as giving off the same light as sixty candles — but it was still work and still fussy. Former president Jimmy Carter wrote:

> Our artificial light came from kerosene lamps, and it was considered almost sinful to leave one burning in an unoccupied room. The only exception was in the front living room, where we had an Aladdin lamp about five feet high whose asbestos wick miraculously provided illumination bright enough for reading in a wide area. We turned this flame way down when we went to eat a meal, both to conserve fuel and to avoid the lamp's tendency to flame up and blacken the fragile wick with thick soot. When this happened — a mishap for which someone always had to be identified as the culprit — we had to endure an extended period of careful flame control while we waited in near darkness for the soot to burn off enough for us to read again.

And kerosene lamps carried another ancient danger for mothers: "You know, you couldn't leave a baby that could move around at all in a room with a lamp or a candle. So you either had to keep the baby in the dark or stay there with him."

There is a difference between living such a life before the development of electricity and living such a life because you are deprived of it. By the 1920s, farmers knew full well that their isolation existed in relation to another world. Yokels, hayseeds, and cabbageheads, they were called. Some promoters of rural life, as much as they understood the necessity of the expansion of electric lines into the countryside, were apprehensive about the changes in expectations it might bring. They saw electricity as part of "this jazz-industrial age [full of] white-lighters, never satisfied, but excited, [who] just don't want to get away from the white lights, out close to themselves, more or less alone." But most farm families had no such apprehensions and grew to resent the absence of electricity — an absence that was obvious every time they visited a city or heard from friends and relatives living in one. Rural free delivery brought catalogs and magazines to their homes, and with them ads for electric irons and washers as well as electric lights. Although electricity still could not be explained, the electric life was idealized — electricity as the good fairy. The women in the ads were washed and made-up, and they wore the fashions of the day, complete with earrings and high heels. They stood as straight as dancers as they delicately pushed a vacuum cleaner with one hand. The modern kitchens in the ads were entirely free of clutter and were white, brilliant white: the enamel electric stovetops and ovens and the built-in cabinets gleamed — no trace anywhere of the soot and ash that farmwomen struggled with daily.

But it was about more than ease and cleanliness. General Electric ads equated the electric life with being a good wife and mother. An advertisement from 1925 declared: "This is the test of a successful mother — she puts first things first. She does not give to sweeping the time that belongs to her children. . . . She does not rob the evening hours of their comfort because her home is dark. To light a room splendidly, according to modern standards, costs less than 5 cents an hour. . . . Certainly no household drudgery should distract her, for this can be done by electricity at a cost of a few cents an hour."

The desire for such modern things meant nothing to a farm-woman. Even if they could be acquired, without central station power they were useless. This was a new kind of isolation.

Perhaps it wasn't the things themselves that many women most desired, but the free time. "The thing [the farm woman] needs in this day and time is electricity. Then when her house is lighted, her cream separated and churned, her washing, ironing, and sweeping, her sewing machine run by the same power, and she relieved from the drudgery of washing and filling lamps, lifting and washing jars, pans, and all these other hard old things, she can have some time for a social life and the improvement of her mind," commented one farm wife. Another said, "We would like to have a chance to live as the city sisters, and not be made to live as a peasant or slave." For both men and women, above and beyond the farm work, they desired simply to be included. This was especially true of the young, who would claim that "everything had already happened before we found out about it" and that "we were back in the woods compared to the rest of the world."

Whether in Texas, Pennsylvania, Iowa, Maine, Alabama, or Colorado, for those waiting for electricity through the twenties, thirties, and forties, there was only time, and waiting for time after. Or perhaps in the later years, the voices hardened even more, for time *had* passed, and as daylight waned and the last chores were done, the life that had spread across fields and woods during the day drew inside. Dark staked its boundary, large and elemental. Families gathered around the kerosene lamp on the table. What had once seemed "the kind of oil people had dreamed about for centuries" had become the symbol of obsolescence and of isolation from the future. "Kerosene light," James Agee wrote in *Let Us Now Praise Famous Men*, "is to electric services what foot and mule travel is to travel by auto and airplane, or what plowed clay is to pavement, and . . . these daily facts and gulfs have incalculably powerful and in many respects disadvantageous influences upon the mind and body."

The tenant farmers Agee wrote about found a use for the discards of electricity long before the lines came through. In a country burial ground in Alabama, he saw graves of mounded earth marked with pine headboards that would weather away in time. The decorations on and around the graves would out-last the soft wood, however. Some were bordered with white clamshells. Others — women's graves — were decorated with plates, butter dishes, and baskets made of milk glass. And still others were marked with what the people had never had in life. "A blown-out electric bulb is screwed into the clay at the exact center [of one grave]," Agee wrote. "On another, on the slope of clay just in front of the headboard, its feet next the board, is a horseshoe; and at its center a blown bulb is stood upright. On two or three others there are insulators of blue-green glass."

The slow and halting extension of electric lines into the countryside was not inevitable. Historian David Nye notes that "street lighting in the United States quickly developed far beyond functional necessity to include advertising and public-relations spectacles. In contrast, in Scandinavia, Germany, and Holland, spectacular lighting developed more slowly, but the electrification of every home was considered a desirable political goal, and had been 90 percent achieved before 1930." In countries where government regarded the establishment of electric power as a social and political responsibility — and took an active role — rural electrification often developed much more quickly, although government interest couldn't ensure success without the money and infrastructure to develop long-distance lines.

When Vladimir Ilyich Lenin created a long-range plan for Soviet national and economic recovery after years of revolution and war, electrification was central to the plan. It would, Russian Marxists believed, "provide a link between town and country, [which would] make it possible to raise the level of culture in the countryside and to overcome, even in the most remote corners of the land, backwardness, ignorance, poverty, disease, and barbarism." But in 1920, not even the cities in Russia had much of an electric infrastructure. When electrical engineer Gleb M. Krzhizhanovskii "displayed an illuminated map of a future electrified Russia to convince the 8th Congress of Soviets to approve a plan for state electrification . . . Moscow's generating capacity was so low . . . that lighting the bulbs on the map resulted in blacking out parts of the city." Without funding for the infrastructure across the vast country, rural electrification fell far short of Lenin's hopes, though the electric bulb came to be called the "Ilyich light" and stood as a symbol of modernization in Soviet propaganda.

But there were many successful instances of countries developing rural electrification programs. In 1924 Harold Evans, counsel to the Rural Electric Committee of the Pennsylvania Council of Agricultural Organizations, published a survey of rural electrification throughout the world. He described the manner in which Sweden, France, Holland, New Zealand, Canada, and other countries had progressed. Sweden, for instance, had been cut off from coal and oil supplies during World War I and had immediately turned to the production of electricity for power. "Ten years ago, rural electrification was practically unknown in Sweden," Evans wrote. "Today 40 per cent of the 9½ million acres of tilled land has access to electric power. . . . This rapid development has come about through many different agencies, the most important of which are the state owned electric systems in central Sweden, the larger private power companies and the farmers' cooperative societies."

Canada began harnessing the hydroelectric power of Niagara Falls in 1910. The government controlled much of the power, and by 1911 the province of Ontario decided to make the delivery of affordable electricity to the countryside a priority. Although parts of Ontario, which covers more than 400,000 square miles, like parts of the rural United States, would not see electric lines until after the Great Depression and World War II, government intention ensured that most of the rural countryside was electrified much earlier than rural areas in the United States. Evans noted that "the kw.h. production of electricity per unit of population in Ontario is more than twice as great as in the United States and is increasing much faster in Ontario than it is in this country." Although only about 3 percent of American farms were connected to central stations in 1924, he remained hopeful that half of the farms in the United States would be electrified by the early 1930s.

In reality, when Thomas Edison died in October 1931, less than 10 percent of the farms in America were connected to central station power. As he was buried at dusk on October 21 — fifty-two years to the day after the first successful experiment with an incandescent bulb at Menlo Park — his widow, the *New York Times* reported, could see from the cemetery in West Orange, New Jersey, "far off above Manhattan, the sky-glow from the lights his genius gave to the world." As a tribute to Edison, President Herbert Hoover requested that at 10:00 P.M. Pacific time — the hour when the sun would have set over the entire country — the nation turn off all its lights simultaneously for one minute and plunge itself into darkness. Radio stations across the nation would announce the moment. "Mr. Hoover left it to each individual citizen to participate in the minute of darkness, pointing out that if the generation of electric current were halted even for an instant, it might cause death somewhere in the country. 'This demonstration of the dependence of the country upon electrical current for its life and health,' the President declared, 'is in itself a monument to Mr. Edison's genius.'" Most farm families didn't have a radio to hear the announcement, and they would already be in the half dark anyway, gathered around their kerosene lamps.

13

Rural Electrification

In 1908 President Theodore Roosevelt appointed the Country Life Commission to investigate the deteriorating quality of life in rural America. When the commission published its report, it concluded, "It is more important that small power be developed on the farms of the United States than that we harness Niagara." But not until the 1920s, when Pennsylvania governor Gifford Pinchot undertook the Giant Power Survey, did any government agency, federal or state, look extensively at what electrification could mean for rural America. The survey suggested that rural electrification would make it possible for factories to move out of the center of cities and not only relieve overburdened, overcrowded urban areas and the people in them but also modernize rural life and "drive a wedge between women and drudgery." And, the survey said, it could do all this cheaply and cleanly. Pinchot proposed an intensive plan, largely based on coal-fired generating plants, to electrify rural parts of his state, but the Pennsylvania legislature, under pressure from utility companies, failed to approve the plan. Only with the implementation of President Franklin Roosevelt's New Deal in the 1930s did widespread rural electrification begin to become a reality.

Modernization of the countryside had been a concern of Roosevelt's during his tenure as governor of New York, when he'd become aware of the inequalities in electric power distribution at his rural retreat in Warm Springs, Georgia. "When the first-of-the-month bill came in for electric light for my little cottage," he recalled, "I found the charge was 18 cents a kilowatt-hour, about four times as much as I paid in Hyde Park, New York. That started my long study of proper public utility charges for electric current and the whole subject of getting electricity into farm homes."

Roosevelt created the Tennessee Valley Authority (TVA) in 1933 as part of the blitz of legislation aimed at alleviating the effects of the Great Depression. The TVA oversaw development along the Tennessee River and its tributaries, which drained all of Tennessee and parts of Kentucky, Virginia, North Carolina, Georgia, Alabama, and Mississippi. These were some of the poorest rural regions of the country, where the soils had been depleted and eroded by extensive cotton production, careless farming practices, and overlogging; where frequent floods silted the waterways; and where almost no rural communities or farms had electricity. Essential to the TVA's plan was a series of dams and reservoirs on the Tennessee and its tributaries for flood control, navigation, irrigation, recreation, and the production of hydropower for electrification. States, counties, municipalities, and farmers' cooperatives would have first access to the power produced. The utility companies of the region voiced strong opposition to the projects and initiated numerous lawsuits, charging that it was unconstitutional for the government to compete with them directly by selling electric power, but the Supreme Court ultimately upheld the constitutionality of TVA projects.

The TVA undertook extensive regional planning, for

Roosevelt believed that only an integrated approach to the conditions in the watershed could permanently improve the quality of life for the valley's residents. He saw electricity not only as something that would modernize people's lives but also as a moral force capable of improving their sense of citizenship and strengthening ties within the community. "Power is really a secondary matter," Roosevelt insisted.

> What we are doing there is taking a watershed with about three and a half million people in it, almost all of them rural, and we are trying to make a different type of citizen out of them. . . . Do you remember that drive over to Wheeler Dam the other day? You went through a county of Alabama where the standards of education are lower than almost any other county in the United States. . . . They have never had a chance. All you had to do was look at the houses in which they lived. . . . So T.V.A. is primarily intended to change and to improve the standards of living of the people of that valley. . . . If you can get cheap power to those people, you hasten the process of raising the standard of living.

The TVA established experimental and demonstration farms and created, through university experiment stations and extension services, a means of informing farmers of good agricultural practices, such as the correct application of quality fertilizer, the terracing of hillsides to prevent erosion, and the planting of cover crops. It developed programs to inform women on the farm about nutrition, the safe handling of food, and sanitation. It also encouraged the formation of electric co-ops: farmers in a given area would join together to pay for the extension of the lines to their farms. All members of the co-op, whether they lived hard by the power station or at the far end of a rural line, would pay the same rate for service. When in June 1934 the first rural electric co-op in the TVA area — the

Alcorn County Electric Power Association in Mississippi —
began operation, Roosevelt said:

> Now the Alcorn County people . . . did a very interesting thing.
> There they had Corinth, which is a good-sized town, and they
> found they could distribute in Corinth — these are not accurate
> figures — they found they could distribute household power at
> about two cents a kilowatt hour. But if they were to run an elec-
> tric line out to a farm, they would have to charge three cents. In
> other words, the farmer would have to pay more. . . . What did
> the Corinth people do? . . . Voluntarily they agreed to take and
> to pay two-and-a-half-cent power which enabled the farmer to
> get two-and-a-half-cent power. That is an extraordinary thing.
> That is community planning.

The TVA's reach was broad, and although it was accepted
more readily in some places than in others, the lure of the
electric life — for those who could afford it — was clearly en-
gaging. One of the authority's directors, David Lilienthal, with
all the idealism of those involved, wrote in his journal in October 1935:

> There must have been ten thousand country people in Fayette-
> ville to attend the opening of the substation near Ardmore. . . .
> I had a chance to walk around the courthouse yard before the
> speaking and see them and overhear them talking. The enthusi-
> asm about the rural electrification program that I noticed par-
> ticularly while I was speaking to this crowd from the bandstand
> in the courthouse yard . . . was really amazing. . . . There is
> somehow a magic about TVA kilowatts. We have really stirred
> the public imagination about electricity.

The building of an extensive system of dams meant that many
towns, settlements, and farms along the rivers would be
drowned. For the first of these dams, the Norris Dam — built

just below the confluence of the Clinch and Powell rivers in eastern Tennessee — the TVA purchased, often by eminent domain, roughly 240 square miles of land in five counties. The valleys were steep, the forests cutover, the fields eroded and depleted by generations of intensive farming. The young had departed years earlier, moving to the cities to find work, because even in the sparsely populated hills, there were too many people for the land to comfortably support. But as the Depression had deepened, these emigrants had trickled back home, and their return had put even more pressure on the land.

Families had lived here for generations, residing in small, isolated communities of farmers and tenant farmers, crossroads stores and churches. Some had never traveled as far as Knoxville. The farmers, mostly self-sufficient, produced a little extra — eggs, butter, vegetables — to barter at the local store for coffee, salt, flour, and plowpoints. Such stores would be "full even without customers," Eleanor Buckles wrote,

> with boxes and bins and barrels and the home-made furniture, baskets, weaving and carving and fox skins brought in for barter, all crowded together. Boxes of shoes and sacks of feed and bolts of cloth overflowed from the shelves and cluttered the floor. In the rear stood the barber-chair circled by boxes and barrels for seating. . . . The gasoline lamp in the center of the ceiling shed a glaring white light and patterned the walls with the shadows of chains and harness hanging from the dusty beams.

The people had created an irreplaceable system of interdependence among their independent selves. They helped each other care for the sick and rang the death bell for their neighbors.

> And since there wasn't no communication, people'd hear that bell, oh, for miles around. The way they rang the death bell

was different from any other. They'd pull the cord — the rope — down and hold it for a few seconds and then let it go back instead of letting it ring the natural ring. Everybody'd recognize the death bell, and they knew there was somebody in the community who was dead. Of course the whole community would come in and prepare food and help and do anything that was needed to be done for the family.

More than three thousand families were forced to leave their land in the Norris basin. The TVA offered "market value" for the properties, but no one felt it was enough — and in truth it wouldn't be: the community would be scattered; the world they moved to would be nothing but strange. Most, in the end, went only with the greatest reluctance. John Rice Irwin, who was a child when the Norris basin was flooded, remembered:

I guess they felt that they were doing it for the benefit of their area. . . . And they especially felt this later on, I believe, when they saw what TVA had accomplished. I think it was somewhat similar to a person going into the army, in the past, you know. They didn't want to go, they dreaded to go, and it was disruptive, but at the same time they felt some obligations. . . . It's very difficult to describe the attachments that they had for their land, their emotional involvement, and the fact they were going to have to leave all that and come somewhere else. It wasn't just that they had spent their lives there, you know, but as far back as their grandparents could remember.

Before the reservoir was filled, more outsiders than ever arrived — engineers, relocators, writers, photographers. Lewis Hine created enduring images of life there before the flood, of women washing clothes in the yard in their zinc tubs, of children sitting obediently in rows in the cabin schools. "And the people up there felt that they were being portrayed as if they were isolated, ignorant, mountain people," Irwin noted. "And I don't know what part TVA played, whether the pictures were from TVA, or whether it happened at the same time; but that was the one big criticism that I recall more than anything else, I think."

Anything made of material that might float to the surface and clog the dam — wooden walls and roofs, tin — was demolished or hauled away. Chimneys and concrete and stone were left as is; the water could just rise over them. The three thousand families had five thousand dead who were exhumed and reinterred on higher ground. When the water rose, the best soil — the bottom soil — was inundated along with the dusty yards and the woodlands and creeks. Water crept along the ridges and flowed into the hollows, and the roads that led into the valley from then on led to calm water.

For all the TVA's original intentions, there was no real plan to fully resettle the people. Most dispersed across the county to land that was just as marginal as what they'd been forced to leave. It was probably easier on the young than on the old, who had little chance of adapting to a new community so late in their lives.

The TVA did build a town about twenty miles from Knoxville. Norris, Tennessee, was influenced by the garden city movement of late-nineteenth-century England, which attempted to humanize the industrial city by promulgating the creation of modest, walkable, self-contained towns enfolded in protective greenery. In Norris, every fully electrified cedar-shingled house had a porch facing its neighbor and was within walking distance of stores, churches, the post office, and other services. The town was ringed with woodlands. Those who lived there, it was imagined, would have extensive opportunity to study agriculture, the arts, and trades.

But like the TVA itself, the vision of Norris was one thing, the reality another. The first buildings constructed served as dam workers' dormitories, and the homes that were built later were occupied by professionals involved in dam construction. Norris never housed the dispossessed. Almost no local families — neither former landowners nor tenant farmers — settled in Norris, and no blacks were allowed. "From all this, the Negro . . . is to be absolutely excluded," wrote Cranston Clayton.

> He cannot even live on the outskirts of the town in his customary hovel. . . . Southern towns will at least allow their out-caste population to live in dirt and shacks down by the creek or the railroad track. But the government does worse. It absolutely excludes them. This blow is all the more disheartening because it is delivered by the United States government. The Negro looks to the government as his best if not his only friend. . . . Federal Courts have been about the only agency by which Negroes felt they could protect themselves as American citizens. . . . Norris is built on government property. The project is nationally supported and therefore ought to be somewhat independent of local prejudices.

The National Association for the Advancement of Colored People (NAACP) undertook repeated investigations of the TVA, charging that it engaged in discrimination in the hiring and housing of blacks. When the NAACP published its findings, the TVA answered the charges by insisting that it could not find enough skilled black labor to fill the positions. The discrimination in Norris was never rectified. In the end, it would become a bedroom community for Knoxville.

As for electrification of the area surrounding the Norris

Dam, many of the relocated were reluctant or unable to electrify their homes, and most wouldn't have electricity until after
World War II. "A malaria-ridden, poverty-stricken, one-crop
population farming burnt-out land can't buy electricity," wrote
Buckles, "nor can it buy the products of factories."

The Rural Electrification Administration (REA), established
by the Roosevelt administration in 1935, two years after the
TVA was under way, didn't directly involve itself in social engineering; it had the more straightforward mission of delivering electricity to rural people across the country. "We were
all feeling our way along," recalled the second administrator
of the REA, John Carmody. Morris Cooke, Carmody's prede-
cessor, envisioned that the REA would distribute low-interest
government loans directly to power companies. With the
money, the utilities would extend their lines to provide widespread electric power to the countryside, and in return for
the favorable interest rate, they would reduce their excessively
high charges to rural customers.

But the private utility companies were still unable to see the
potential in farm kilowatts, especially in the precarious economic years of the early 1930s. By that time, most of the electric utilities in the United States were inextricably bound
up with large holding companies — a model initiated decades
earlier by Samuel Insull, who, while bringing electricity to
suburban Chicago neighborhoods, also systematically bought
up controlling interests in the small, outlying utility companies in the area and combined them with other assets. Holding
companies were attracted to the stability of the utility companies, which they could use to guarantee other, often riskier investments, but such a practice linked the financial security of
utilities with these other investments. Not only were the large

holding companies more volatile than separate utilities, but their reach also extended across large geographical areas.

During the stock market crash of 1929, the enormous losses suffered by large holding companies compromised the financial health of utilities. Economically fragile utility companies were not just a liability for stockholders. Since the utilities were now less creditworthy, it cost them more to borrow money, and this cost was passed on to consumers. In 1935, in an effort to bring stability and control to the utility industry, President Roosevelt signed the Public Utility Holding Company Act (PUHCA), which strictly regulated the size and type of companies that could hold stock in utilities. Among other things, the legislation limited the amount of debt such companies could accrue, allowed the government to set electricity rates, and mandated that utilities sell power to everyone in exchange for being granted exclusive control over a given service area.

Even with such regulations in place, utility companies failed to extend electric service to rural parts of the country, so Cooke began a program that established rural cooperatives like that in Alcorn County and other TVA communities. Farmers and farm wives ran the co-ops together. Members kept the books, read their own meters, and engaged in troubleshooting when things went wrong. The REA loaned rural districts money not only for lines but also for wiring of individual houses, and the Roosevelt administration, knowing it wouldn't be feasible to extend the lines for light alone, created a federal credit agency, the Electric Home and Farm Authority, which subsidized the purchase of refrigerators, stoves, and hot water heaters, all of which would increase household electricity usage at the same time it modernized rural living. Where feasible, communities might construct small electric plants, but most of the time they purchased power wholesale from existing utility companies.

By 1938 the REA had financed about 350 projects in 45 states. "Initially . . . the REA benefited a relatively small group of people — primarily those farm families in the middling ranks . . . and those who lived in rural areas that had a critical population mass," notes historian Katherine Jellison. It would be decades before the most isolated and poorest communities would see power lines come through. A co-op might encompass several small towns and include stores, Grange halls, gas stations, schools, and other town buildings as well as the surrounding farms. Typically, a co-op constructed more than two hundred miles of lines, for which it borrowed about a quarter of a million dollars from the REA.

To conserve funds, the span lengths were longer, which meant there were fewer poles per mile than in urban centers. To protect the cables from strong winds and icing, they were reinforced with steel. Initially, a mile of rural line cost $2,000 to construct, but this soon dropped to about $600, in part because the work became more efficient. Lines were strung by waves of crews: one mapped out the project, another dug holes, another erected poles, another played out the line, and so on. You can see the linemen in old photos — in the back of a truck squatting among rolls of wire, harnessed to towering poles, and walking alongside horses drawing poles. And because rural people had been waiting for decades for something they felt had been denied them, they often thought of the linemen as heroic. One account notes, "Construction crews . . . have dug post holes in ground frozen 3 feet down. They have set poles when the snow was waist deep." Another reports, "An Indiana woman lay dying of pneumonia in her farmhouse.

The doctor said that an oxygen tent might save her, but there was no electricity in the house to operate the tent fan. Three linemen, working in a driving rainstorm, built a 500-foot extension in just two hours. The switch was turned on and the woman's life was saved." One legendary crew outside Kansas City, Missouri, was known as "the Four Horsemen of the Lines." The poles themselves — slender, usually with just one crossarm — were called "liberty poles."

The utilities quickly understood that they had underestimated the needs and desires of many rural communities, and in an effort to subvert the success of the co-ops, some attempted to skim the most lucrative customers — those living nearest towns and those who were most prosperous — for themselves. Just prior to co-op lines going in, a regional power company would put up poles — even in the middle of the night — to siphon off these customers. Spite lines, they were called, or snake lines, for they almost never ran straight but crisscrossed an area. One REA cooperative specialist recalled: "In Virginia, a co-op engineered a line north through the wilderness, ending in a prosperous dairy section near Chancellorsville. When construction was about to start, the power company built a short line out of Chancellorsville to serve a handful of the large-consumption dairies on which the co-op had counted to makes its 40 miles of line feasible." Such tactics, of which there are more than two hundred recorded cases, could weaken a co-op's effectiveness and ruin its chances to prosper.

By the time electricity came to the country, light bulbs were brighter, washing machines more efficient, and irons more streamlined. Farm people who could afford it bought multiple appliances before their homes were even hooked up to power, or they got appliances secondhand from city friends, so unlike in the early years, many experienced the full gamut of electricity all at once. Their kitchens, no longer cluttered with gray zinc tubs and pails, with washboards and wood stoves, were bright with white enamel stoves, refrigerators, and washing machines. Their homes were filled with little whirs, buzzes, and hums. One woman, about two years after she was married, recalled:

> I had gotten these beautiful wedding presents. An electric coffee maker, an electric toaster, and there they sat. . . . So the day the electric came in, I sat at my kitchen table. The electric coffee pot was plugged in, the toaster was plugged in, a bare light bulb hung above, and I sat there and waited. . . . And such a thrill, you have no idea. I had polished all my oil lamp globes. They were sitting in a nice neat little row. Never again would I have to polish those sooty old things. Never again would I have to fill the tank on them, never again would I have to trim the wicks. They sat there and I was glad.

Those who'd had battery-run radios before line electricity had had to mete out the listening time: "We had a large battery-powered radio in the front room that we used sparingly, and only at night, as we all sat around looking at it during 'Amos and Andy,' 'Fibber McGee and Molly,' 'Jack Benny,' or 'Little Orphan Annie,'" former president Jimmy Carter remembered. "When its power failed we would sometimes bring in the battery from the pickup truck to keep it playing for a special event." In the fully electric life, however, the sounds of other voices, music, applause, jokes, the weather, and farm reports could fill the air all the time. "The day we got our radio we put it in the kitchen window, aimed it out at the fields, and turned it on full blast," one woman recalled. "During the first week, the workers hated to be out of the sound of it." But there were also mild complaints from farm wives:

> They report that their husbands are spending more time than ever in the barns experimenting with their electric milkers and coolers. A lot of men have put radios in their barns — for their own amusement, their women folks think — but the men tell me their cows give more milk to the strains of music than without it. . . . As a result of this modernization the wives of these farmers tell me that for the first time it is hard to get the men in to meals. They act like boys with new tool kits, always puttering around with new equipment.

For Carter's family and their neighbors, electricity changed their sense of themselves and their community. "I think the best day of my life — the one I remember most vividly with the possible exception of my wedding day — was the night they turned on the lights in our house," he recalled. "Also the bringing of the rural electric program to the farms of our Nation made it possible for us to stretch our hearts and stretch our minds to encompass public involvement in affairs that would not have been possible without the rural electric program." One Pennsylvania farmer remarked, "We felt like first-class American citizens." Another said, "Electricity changed the country way of living. That was the beginning of the change, right there. It put the country people more on a par with the city people."

Light may have been the least of it; certainly electric irons, washers, pumps, and milking machines would make a greater difference in their lives. But in the late 1930s and 1940s, when electricity finally came, it was the light they were waiting for. To see (and be seen) beyond the circumference of the kitchen table, to see into the corners of a room or into a husband's face in the evening, "was wonderful. Just like going from darkness into daylight." One farmer observed, "I'll never forget the day when they announced the electric was turned on. I waited till dark to do my chores. I had the barn all lit up like a Christmas tree. Oh, that seemed nice, especially the stable — you didn't have to look where you was goin'." The moment a house was supposed to be connected to the electric lines was known as "zero hour," and people would flip their switches on and off to make sure they didn't miss the instant of connection. The first thing some did once they were hooked up was to turn on every light and then drive down the road just to look back at their illuminated home.

For those in cities, electric light already possessed the bone-weary, jaded cast of Edward Hopper's diner in the small hours: the attendant, the couple, the lone man trapped within. How they arrived or how they will leave is a mystery. At the same time, rural men and women stood bewildered before the one bare bulb hanging from their kitchen ceilings. Some screwed corncobs into light sockets to keep "the juice" from leaking out, or they would not let go of the chain pull, believing that once they released it, the light would go out.

Sometimes people — parsimonious farm families — kept their lights on all night long: "That light in the kitchen came on, and that was the prettiest sight I ever saw. It was wonderful after all those years of oil lamps. I never expected to get it, unless I went away from here." And it was the light the line-men remembered. "Some of them wanted you to come and turn it on for them," one recalled. "They was a little afraid you know. They didn't know anything about it. So you go, there's nothin' to it, just turn the switch on and you've got it, see. And so I turned the light on, oh my gosh, look at that. We've never had it that way — you can see all around the room." Another said, "I've seen this happen — the lights come on — hundreds of places, and its an emotional situation you can't describe. . . . Something happens, lightning strikes them and they all at once are different. People prayed, they cried, they swore."

What of kerosene, which for a brief time had seemed the democratic perfection of light? In memory, some children will fondly recall the oil lamp in the kitchen after supper or the lantern moving across the yard as their father returned from his chores, but few wish to return to those days. When one co-op in Pennsylvania finally strung their lines, they held a mock funeral for a kerosene lamp: "Buried here May 3, 1941 by the Adams Electric Cooperative as a symbol of the drudgery and toil which its member-families bore far longer than was necessary or right but which, with the energization of their own power system, are now abolished for all time." Mock funerals were held in other communities as well. Elsewhere, farmers and their wives were content simply to smash their lanterns on the ground.

Rural people were used to being self-sufficient — repairing their own plows, saving their own seed — but electricity was a mystery, and electricity manuals for farmers reflected the old bewilderment: "What is electricity? . . . No one today knows the exact answer. All that is actually known is that this powerful energy is present in the world, and that it has been 'harnessed' so it can be used as a safe, tireless and efficient servant of mankind." And now, like city people, they were tied to a vast network. When a quiet winter rain fell and the temperature dropped and ice built up on the wires — and on the tree limbs hanging over the wires — they'd hear the sound of cracking, like rifle shots, and catch the scent of pine, then darkness would overcome them once again. Their electric milking machines stood useless in the pitch-black barn; the heat was gone in the chicken coops and incubators. As one farmer observed, "All this pushbutton stuff. Well, it becomes a part of you. You can't cook a meal without it; you can't take a bath without it; you can't get a drink of water without it. . . . There you are, you're hooked. . . . [Before if] you had an Aladdin lamp you could light it and have a good light and go right on about your business, see, but you're hooked when the power goes off."

Electricity meant that the children of farmers would be different people. Not only would they do better in school once they began studying by electric light, but it would carry them into a different world: "To a farm girl who has been brought up with many electrical conveniences it is like listening to a fairy tale to be told that once rural homes did not have electricity."

Sometimes electricity did give a farm more possibilities. "I would never have believed what it has meant," said one farmer. "My boys who are just entering or about ready for high school are making their plans already about what they are going to do, in the country, when they grow up. It used to be they talked about what they were going to do when they grew up, seeming to have in mind everything else except farming." But it couldn't entirely staunch the departures: the number of farms and farm families continued to decline. Most rural children vanished into the glare of the modern world.

But the "liberty poles," it turned out, worked both ways. The extension of electricity into rural areas spurred the movement of city people to the countryside, bringing the "white-lighters" to the farmers' doors. The advancing electric lines, says sculptor John Bisbee, were like ferns uncurling, or so it seems in three aerial photographs of Dunbar Hill in Waitsfield, Vermont, the location of Bisbee's family farm. The pho-

tograph from the 1940s captures a world on the cusp of electrification: one simple road, three farmsteads. In the photo
from the 1950s, the lines have made their way down the main
road, and side roads — like nubby furled leaflets — are beginning to sprout on either side of the main. In the last photo,
from the 1960s, those roads have opened further into the old
wild, and farther along the main road are yet more nubs. To
Bisbee, in the latest aerial shot, the houses and their clearings
seem to shine from out of the wooded dark.

14

COLD LIGHT

Practically every illuminant in use to-day is patterned after
the sun and stars. . . . No artificial lamp is known but that gives
off ample heat to be felt by the hand. It is all "hot light."

—E. NEWTON HARVEY, 1931

OVER DECADES, INCANDESCENT BULBS had grown
far stronger and more dependable than those first as-
sembled in Edison's factories. The quality and strength
of the glass had improved, as had the efficiency of the vacuum.
Most important, the filament had evolved from carbon to
tungsten and finally ductile tungsten (tungsten alone is quite
brittle and therefore fragile). By 1922 renowned General Electric scientist Charles Steinmetz could claim, "Today we are
producing . . . sixty-eight times as much light as we could produce with the lights in use fifteen years ago." The greater brilliance required greater heat, of course, and ductile tungsten
filaments are hot: "A 60-watt bulb operates at a temperature
twice as high as that of molten steel in a blast furnace. Asbestos or fire brick would melt like wax at such a heat. Yet the tiny
filament wire in the lamp measures less than 2/1,000 inch in

diameter — finer than a human hair." While such heat had its practical uses — to incubate chicks and keep piglets warm — in homes, offices, and factories, it largely went to waste. This was acknowledged even by Tesla, Edison, and others in the incipient years of incandescence. As early as 1894, one *New York Times* reporter exclaimed, "What a preposterous dissipation there must be of the energy stored in a lump of coal between its first liberation by combustion and its final emergence in the form of electric light!"

By the 1930s, coal powered much of the growing electric grid, and government officials had become concerned about the stress the ever-increasing use of electricity was exerting on known coal reserves. Additionally, labor strife in the mines sometimes affected the supply of fuel to power stations, so the development of a less wasteful illuminant — a practical "cold light" — had great appeal. Toward such an end, physicist E. Newton Harvey undertook extensive studies of bioluminescence in the natural world — glowworms, the gills of mushrooms, jellyfish, foxfire, beetles, fireflies — in an attempt to reproduce its effects for practical human light. Harvey had great hopes for bioluminescence because the reaction between the chemical compound luciferin and the enzyme luciferase, which produces bioluminescence, is extremely efficient: virtually all the energy generated goes toward creating light; almost none is lost as heat. Additionally, the reaction is reversible. As Harvey noted, "Here you have an animal that makes its fuel and burns it and produces light . . . and then it takes the combustion product and reconverts it into fuel again, and the fuel is ready to be burned a second time. The firefly is able to unburn its candle."

Humans have historically used bioluminescence to see in the dark, and not only as a last resort, the way pitmen used glowing, rotting fish to work in the fiery Tyne mines. For centuries throughout Southeast Asia, people gathered fireflies and released them into tight wooden cages or perforated, hollowed-out gourds so as to have light in the evening. Sometimes they let them loose into the trees to illuminate tea gardens and pathways. In nineteenth-century Japan, capturing fireflies was a gainful means of employment:

> At sunset the firefly hunter starts forth with a long bamboo pole and a bag of mosquito netting. On reaching a suitable growth of willows near water he makes ready his net and strikes the branches twinkling with the insects with his pole. This jars them to the ground where they are easily gathered up. . . . But this must be done very rapidly, before they recover themselves enough to fly. . . . His work lasts till about 2 o'clock in the morning, when the insects leave the trees for the dewy soil. He then changes his method. He brushes the surface of the ground with a light broom to startle the insects into light; then he gathers them as before. An expert has been known to gather 3,000 in one night.

Even a few fireflies might provide enough light by which to see. At the end of the nineteenth century, in the Smithsonian Institution light collection, there was a dark lantern said to have been used by a thief in Java. The shallow wooden bowl had been fashioned with a pivoting lid that could be used to hide the light in a hurry. The thief lined the cup of the lantern with pitch and stuck several fireflies to it. When one firefly perished, he replaced it with another from a store he kept in a capped cane stalk.

In the southern regions of the Western Hemisphere, people sometimes saw by the glow of a bioluminescent click beetle,

Pyrophorous noctilucus, which emits a constant green light. A history of Hispaniola written in 1725 attests:

> There were at first found a sort of vermin, like great beetles, somewhat smaller than sparrows, having two stars close by their eyes and two more under their wings, which gave so great a light that by it they could spin, weave, write, and paint; and the Spaniards went by night to hunt the Utias, or little rabbits of that country . . . carrying those animals tied to their great toes or thumbs. . . . They took [the beetles] in the night with fire-brands because they made to the light and came when called by their name, and they are so unwieldly [*sic*] that when they fall they can not rise again; and the men stroaking [*sic*] their faces and hands with a sort of moisture that is in those stars, seemed to be afire as long as it lasted.

These beetles are the brightest of all luminous insects — the Spanish conquistador Bernal Díaz del Castillo, thought a flurry of them were the matchlocks of his enemies — and during nights of almost complete darkness, the beetles in any number must have seemed magical and spectacular, though they are actually less than two inches long (nowhere near the size of sparrows), and few of us today would think them bright enough to help us work or walk.

Steinmetz had put great store in Harvey's work on biolumi-nescence: "I think it is possible that twenty years from now it may be a thing of tremendous practical importance. . . . There is, of course, no absolutely cold light, but there are experiments on many which may be called comparatively cold. . . . There are none, however, which compare with that of Dr. Harvey, in its promise of working at low cost. All other kinds require coal or energy of some other kind to produce electri-cal power." Throughout his decades of research, Harvey succeeded in understanding more clearly the way bioluminescence works, and he was even able to diffuse enough luciferin in a flask of water to create a light steady enough to read a newspaper by. But neither he nor anyone else managed to turn it into a practical light for industrial society.

The nearest researchers came to cold light in the 1930s was the fluorescent tube, which uses about a quarter of the energy and emits a quarter of the heat of incandescent bulbs of the same strength. It's a descendant of nineteenth-century discharge lamps, which used various gases and combinations of gases to create different-colored lights: neon for red, argon for lavender, mercury and argon together for blue, and helium for yellow. All such lights eventually came to be popularly known as "neon lights," and although they proved to be ideal for signs and advertising, researchers were unable to find a gas alone or in combination that could produce a practical white light for workplaces or homes. Peter Cooper Hewitt came closest, just after the turn of the twentieth century. He fabricated a mercury vapor lamp — a four-foot-long tube that shone greenish blue — which could illuminate outdoor spaces and had some industrial applications, but its size and strange hue weren't fit for interiors.

Fluorescent light — developed between 1934 and 1938 at the General Electric laboratories in New York — unlike earlier discharge lamps in which the gas itself was an illuminant, requires a second conversion. For this purpose, the glass tube, which contains mercury and argon, is coated with a phosphor on the inside. An electric current vaporizes the mercury (the argon helps to start the electric arc), and the mercury gas then transports the current through the tube. As it does so, it produces ultraviolet light, which is invisible to the human eye. The phosphor coating, however, glows — or fluoresces — in the presence of ultraviolet light and creates the light we see. Different phosphor coatings produce different shades of white, as well as some colors.

Even after researchers produced a technically successful fluorescent light, marketers at General Electric were unsure whether the public would take to something so different from an incandescent bulb. The shades of fluorescent white light all had a colder cast than that of incandescent light. The long tube was not only bulky and distributed light differently, but it also could not simply be plugged into a traditional socket or screwed into an incandescent fixture; it required specific fittings. And fluorescent fixtures would not allow for interchangeable lights: a fixture for a thirteen-inch tube could accommodate only that size light. Most at General Electric thought the fluorescent light would be used largely for decorative purposes, and when the company introduced fluorescent lights to the public at the 1939 New York World's Fair, where they accounted for one-third of all the exterior illumination, they did have a distinctly decorative slant.

The fair rose up out of the swamplands of Flushing Meadows, in the borough of Queens, New York, at a time when the United States was still mired in the Great Depression. Its theme, the World of Tomorrow, aimed to cast an affirmative eye on the future, the future being a 1960 of clean, orderly cities, surrounded by satellite villages — Pleasantvilles, each with a population of ten thousand — interspersed with modern farms (albeit farms where workers walked home in a sentimental dusk shouldering hoes and scythes) and tame, green open spaces. An interstate highway system would safely carry cars traveling a hundred miles an hour across the country, and television — also introduced at the fair — would bring a brave new visual world into homes. It was also a fair inundated with brand names — Eastman Kodak, General Motors, General Electric, Westinghouse — as E. B. White clearly saw:

> The road to Tomorrow leads through the chimney pots of Queens. It is a long familiar journey, through Mulsified Shampoo and Mobilgas, through Bliss Street, Kix, Astring-O-Sol, and the Majestic Auto Seat Covers . . . through Musterole and the delicate pink blossoms on the fruit trees in the ever-hopeful back yards of the populous borough, past Zemo, Alka-Seltzer . . . and the clothes that fly bravely on the line under the trees with the new little green leaves in Queens' incomparable springtime.

By 1939 lighting designers and architects were able to work with a variety of brilliant, durable lights, which they could employ to create natural fadeouts and highlights. Graduated shades and intensity of light created more sophisticated effects than those used at the World's Columbian Exposition in 1893, when architects relied on floodlighting façades or outlining buildings with bulbs that, for all their novelty and brilliance, diminished the apparent size of the buildings at night and muted the details and nuances of their surfaces. The more advanced lighting effects of 1939 not only enhanced and punctuated details of the buildings but also granted structures a distinct appearance at night, completely different from the way they appeared during the day. And architects could now design buildings constructed almost entirely of glass, which not only showed off interiors at night but also made interior light integral to exterior illumination.

One reporter at the fair observed: "Only selected parts of the buildings glow. . . . The solid architectural structure of the daytime is set aside for an immaterial structure of light. It is not the intention that the Fair by night shall be the same Fair that was seen by day. After dark it is changed into a lightscape." Nowhere was this clearer than at the center of the World of Tomorrow, where there stood stark white modernist versions of a spire and a dome: the 610-foot-high, three-sided obelisk called the Trylon and the 180-foot-diameter sphere built of steel and cement stucco called the Perisphere. The sphere's eight supporting steel columns were masked by a ring of fountains, so that from a distance it appeared to be floating on water. The severity of the daytime architecture turned magical in the dark: "As night fell, the globe was bathed in colored lights — first amber, then deep red, and finally an intense blue — on which were superimposed moving white lights (filtered through mica) in irregular patterns." The results, one historian attests, "bore an uncanny resemblance to the views of Earth which would be taken from Apollo spacecraft some thirty years later."

Throughout the grounds, white fluorescent tubes — encircling the midsections of tall flagpoles, cinching them like a belt — lit pathways. Colored fluorescent lights backlit murals, illuminated signs, and highlighted walls. Whether concealed, recessed, or ghosting structural details, they created sleek, striking effects:

> Even the drabbest and most monochromatic of buildings sprang to life under the influence of creative lighting techniques. By day, the only touch of color that relieved the honest metallic finish of the U.S. Steel dome was the minimal application of blue paint to the external ribs that acted as its structural supports. But by night the ribs glowed a bright azure that the shiny steel surface reflected and the entire dome gleamed with a cool radiance. . . . One of the most spectacular applications was the

design of the Petroleum Building, a triangular-plan structure featuring fins of corrugated steel ascending its outer surface in four concave strips. Behind each strip a trough containing blue fluorescent tubes produced indirect illumination that made the building's horizontal segments seem to float independently in space.

The use of fluorescent light in such spectacular ways helped make illumination at the fair a great success, but it didn't settle the questions marketers at General Electric had concerning them. Would people be persuaded to adopt them for the ordinary light of their homes? Fluorescent lights buzzed. They flickered and hummed. There was a delay when you turned them on. They grew dimmer and less efficient over time. Although they eventually gave more light for less cost, above and beyond the special requirements for their installation, they were more expensive to purchase. And they *were* cold: they cast a white light unbecoming to faces and surroundings.

The General Electric advertising campaigns for fluorescent lights emphasized their utility, suggesting how and where to place the bulbs, especially in kitchens — over sinks, stoves, and countertops — to most effectively illuminate tasks and reduce fatigue for the eyes. One ad announced: "It's easy to see into pots and pans. Easy to measure ingredients. Easy to see whether dishes are clean." And what success fluorescent lights had in homes was largely functional. Beyond kitchens, they illuminated bathrooms and work areas in cellars, but few found their way into living rooms and bedrooms.

Still, fluorescent light offered an efficient, economical way to illuminate the large interiors of offices, factories, and department stores, and in the years after the New York World's Fair, they became ubiquitous above assembly lines, in cubicles and doctors' offices, on manufacturing floors, and in warehouses. They even inspired the construction of some windowless factories. Colored fluorescents lit theaters and restaurants and were used for display and advertising. General Electric sold 21 million fluorescent lamps in 1941, and by mid-century more than half the interior lighting in the United States would be fluorescent.

Perhaps their ubiquity in public spaces and workplaces made them seem doubly cold for the home, a place where people often wanted a relaxing, warm interior. But fluorescent light's failure to make domestic inroads could also be testimony to the particular place incandescence held in American life. By the time the World of Tomorrow opened, 90 percent of urban homes in the United States were electrified, and the incandescent bulb had worked its way fully into the imagination. Indeed, its shape floating in a thought bubble had become a metaphor for a bright idea — a tribute both to the revolutionary place of electric light itself and to the genius of Thomas Edison, whom almost everyone perceived as the sole inventor of the bulb. Not only was the race for cold light something of an abstraction to those outside the laboratory, but also nothing about the development of fluorescent light could match the public drama that unfolded at Menlo Park. Incandescent light — clean, bright, economical, instantly available with the flick of a switch — meant so much. Why would people want anything else?

WARTIME: THE RETURN
OF OLD NIGHT

The earth grew spangled with light-signals as each house lit
its star, searching the vastness of the night as a lighthouse
sweeps the sea. Now every place that sheltered human life
was sparkling.

—ANTOINE DE SAINT-EXUPÉRY,
Night Flight

O N SEPTEMBER 1, 1939, while fairgoers in New York
were marveling at the Perisphere, Nazi troops moved
into Poland, and evacuations from London and other
major British cities to the countryside began. At sunset of that
day, the British government issued its first official blackout order. From the heavens, it was hoped, London would appear
little different than an oak forest or a heath and so escape the
fate of the city in the previous war. Across Europe during
World War I, it was by their lights that people were betrayed,
as airmen carried out strategic bombing of cities and towns at
night. They could navigate by tracking human lights, but the
planes attempting to intercept them could do little more than
chase shadows. "Strategic" may be an overstatement, however.

Guidance equipment then was so rudimentary that except on clear nights with a full moon, the bombers often missed their intended marks. "Experience has shown that it is quite easy for five squadrons to set out to bomb a particular target," observed one British bomber pilot, "and for only one of those five ever to reach the objective; while the other four, in the honest belief that they have done so, have bombed four different villages which bore little, if any, resemblance to the one they desired to attack."

It was all so new — the first recorded instance of aerial bombardment dates to 1911, when an Italian pilot lobbed hand grenades over the side of his airplane as he flew over the oases outside Tripoli — that throughout the First World War, European cities had no real defense against attacks from the air. England alone endured about a hundred air raids and suffered more than fourteen hundred casualties from them. Those who survived knew that the next war would be even more perilous — there would be more light, better planes, and more sophisticated guidance systems.

After the end of hostilities in 1918, in addition to concentrating on building up its air fleet and increasing the sophistication of its bombs, the British government intermittently considered how best to protect its urban population from future aerial attacks, though it wasn't until 1936, with the heightened threat from Hitler's Germany, that officials began to formalize plans. Their strategies for survival included the creation of a public warning system, evacuation plans, the construction of shelters, the digging of trenches, and — what would prove to be the most difficult to endure — preparations to hide or douse all artificial lights. The blackout, unlike an air raid warning, would not be intermittent. It would last, to one degree or another, for the duration of the war. Such preparations required years of planning, for all that had advanced from the single lights on the sills of village houses in the seventeenth century — the second and third shifts in factories; the evening hours in shops and stores; the freedom light lent to the hours after dark; all the leisure, buoyancy, energy, and ideas that countered the limits and fear of the old long nights — would have to be concealed.

To cut the main power supply, it was argued, would be such a severe burden on the populace that the British government's first plans involved darkening the city by other means. It was imagined that those in charge of public lighting might keep large staffs on standby who were prepared go through the neighborhoods removing bulbs from streetlamps one by one. Within a single district of London, this could take six hours. (Ultimately, almost all streetlights would be darkened for the duration of the war.) Factories and industrial complexes not only would have to screen windows and eliminate external lights, but they also would have to contend with the glow from the fires in clay works and glassworks, the fires in blast furnaces and coke ovens, and the burning slag heaps. Those in the steel industry estimated that it would take three years to screen their furnaces and that the cost of dealing with coke ovens alone would cost £300,000. Shops and stores would have their hours of business restricted and would not be allowed to illuminate their signs or their plate glass windows. Cinemas and theaters would not only have to darken their marquees; they would have to close: large gatherings, officials feared, would offer bombers an effective target. Churches and cathedrals — built for the glory of light — found it almost impossible to obscure their windows, so they planned to hold their evening services in the afternoon.

Railroad officials confronted ways to cut the light in marshaling yards and in the interiors of trains, to eliminate the arcing of electric trains, and to hide signal and train lights. Headlights on ambulances, trucks, and buses were to be masked down to thin horizontal slits. Civilian cars were to have no light at all, and Britons would depend on white paint on curbs and at intersections to guide them as they drove at night. As for pedestrians, there were to be no flashlights; they would not be allowed even a match to see by. To find their way into their own homes after dark, citizens would have to dab some white paint on the doorknob or bell. Without a moon, for all the brilliance of the winter stars — which would become visible over London as they hadn't been for centuries — they wouldn't be able to see their own hands in front of their faces.

Householders would have to cover their windows with black paint, oilcloth, or blinds made of thick black paper, which would have to be sealed to the window frame — not a sliver of light could leak out. The fines for violations would be stiff, the enforcement rigid.

When the first real blackout was ordered in September 1939 and all these measures went into effect, many people slept with their jewelry and money, a flashlight, and a first-aid kit on a chair by the bed. In addition to covering their windows, the more conscientious taped the glass in hopes of protecting it from shattering and made a room for refuge in the basement, with a strong table to hide under, mattresses, blankets, food, water, candles, and books and cards to pass the time.

To those who ventured into the streets during the blackout, the night world was a wilderness. Not only did they bump into newly placed piles of sandbags and barricades, into barbed wire and machine gun emplacements, but they also walked into familiar walls and trees, into canals, off railway platforms, and into each other. Bus conductors couldn't tell the copper coins from the silver. People didn't always know their own roads or recognize their own houses. Familiars passed each other on the street and traveled next to each other on trains unbeknownst to one another. Without streetlights, and with masked headlights, travel at night became so perilous that in the first four months of the blackout, 2,657 pedestrians were killed in road accidents, twice the number as during the same months the year before.

The first extreme regulations were relaxed in mid-October as a middle ground between safety and fear was negotiated. Blackout hours shrank: the blackout now began a half-hour after sunset and ended a half-hour before sunrise. A little sliver of illumination from streetlights — called "glimmer lighting" or "star-lighting" — was permitted at intersections. Civilian drivers were allowed to use headlight masks. The number of roadway casualties fell, though this was also helped by a new speed limit of 20 miles per hour in densely populated neighborhoods during blackout hours. Gas rationing lessened the flow of traffic as well.

Cinemas and theaters were allowed to reopen, and during the war years the cinema proved to be extremely popular. As long as people sat transfixed in the dark caves of the movie houses, they could forget the lightless world outside and the strain of everyday living — the threat of falling bombs, the rationing of tea and eggs, sugar and meat. As the sprockets of the projector precisely reeled the film forward — one frame moving past the light as another advanced toward it — a shutter closed and then opened again in order to stop the light from projecting during the brief moment the frames were in motion. On the screen, the images moved seamlessly as light and shadows illuminated an upturned face, a calculated act of murder, or a chorus line of dancers. But without those dark moments interspersed between the light, the film would appear to be no more than jerky vertical movements. The illusion of constancy viewers had as they gazed at a man eating his shoe or sliding down a banister in tie and tails would be lost.

At Christmastime, 1939, shops were allowed limited lighting, as were some theaters, on condition that all had to batten down during air raid warnings. Museums and galleries, which had previously been closed, opened, too, though most of their valuable works had been shipped to safer places. Pedestrians were allowed to use flashlights again, but the lamp had to be covered with two thicknesses of white paper and the light had to be switched off during alerts. They couldn't have given off more light than the stone lamps of the Pleistocene.

The distinction between night and day was absolute, as it hadn't been since the shuttered, silent nights of the Middle Ages. Life continued in want and isolation as everyone waited within their husks for peace to favor them again.

By the time of the Blitz, in the late summer of 1940, the Luftwaffe had developed radio beams that helped navigators locate targets, so they could aim with some accuracy in bad weather or on dark nights. Still, the full moon was called the Bomber's Moon. Vera Brittain described the night of September 7, 1940:

From different angles, at different heights and with different speeds, came fifteen hundred aeroplanes of all types and sizes dropping bombs by the ton in eight hours of terror. . . . Furious fires, climbing the midnight sky from slums and docks, destroyed in a moment the simple precaution of the blackout; . . . civilians listening in the shelters and basements to the ceaseless roar of the planes and the intermittent thud of bombs, lost all sense of time, of order, even of consciousness. That night at least four hundred people perished; on the next, two hundred.

October 15 saw 410 raiders over London. They dropped 538 tons of explosives, which killed 400 civilians and injured 900 more. Hundreds of fires burned throughout the city. It would go on night after night for months, then intermittently for years. The two-minute rise and fall of the air raid warning, the sirens through the boroughs, the sound of the raiders coming. "Whatever part of London we live in," Brittain wrote, "they always seem, by day or by night, to be passing just overhead." Then the whistle and crackle of bombs falling, "the clatter of little incendiaries on roofs and pavements"; burglar alarms, dogs barking, glass breaking, fire bells, the rain of rubble, walls tumbling, metal crashing, wood splintering. Sounds — and what the imagination made of them. "Yet another raider came up from the southeast," wrote Graham Greene, "muttering . . . like a witch in a child's dream, 'Where are you? Where are you? Where are you?'"

And what if there happened to be quiet? "Over the night, like a suffocating coverlet, lies that sinister silence, characteristic of all raid periods free from noise, which make us feel that a large number of unpleasant occurrences are happening somewhere else."

At the warning, many left their darkened rooms for the basement; for a corrugated metal Anderson shelter in the garden; for churches, schools, or the tube, where they waited for daylight as the air soured around them. Even deep underground, those seeking refuge could feel the percussive effects of the bombs as they rested their heads against the wall. People were most visible in hiding. "[They] had taken over the Underground. . . . It wasn't only on the platforms it was in an empty tunnel, too, where they were excavating to put a new line in. . . . I had never seen so many reclining figures," remembered sculptor Henry Moore, "and even the train tunnels seemed to be like the holes in my sculpture." He drew them, faces nestled in the valleys of their bedclothes, some with their mouths slack and open, others with their jaws clenched, heads buried in the crook of an arm or turned to another. "And amid the grim tension, I noticed groups of strangers formed together in intimate groups and children asleep within feet of passing trains."

The day was "a pure and curious holiday from fear. . . . The night behind and the night to come met across every noon in an arch of strain. To work or think was to ache." Amid the dust and grit; the stopped clocks and shaken plaster; the vases, toilets, and tables set for dinner exposed to the street; the acrid smell of bombed factories, dye houses, and tanneries; the stink from broken sewer pipes and domestic gas lines — every seeping little loss — funerals crawled along the cobbles and amid the rubble. Summer seeds drifted in and took hold in what remained of kitchens and bedrooms. The dazed walked to their jobs, made their breakfasts, and polished broken glass.

Life was hardly different in Moscow, Berlin, Hamburg, Tokyo, Paris, Dresden, Cologne: all enduring aerial bombardment, all under some kind of blackout order. Stone cities turned into dust. Wood cities burned. Hans Erich Nossack described his approach to Hamburg after a night of bombing:

> What surrounded us did not remind us in any way of what was lost. It had nothing to do with it. It was something else, it was strangeness itself, it was the essentially not possible. . . . And already we were perplexed and did not know how to explain the

strangeness. Where once one's gaze had hit upon the walls of houses, a silent plain now stretched to infinity. . . . Solitary chimneys that grew from the ground like cenotaphs, like Neolithic dolmens or admonishing fingers. . . . How many things we had learned in school, how many books we had read, how many illustrations we had marveled at, but we had never seen a report about anything like this.

The eastern seaboard of the United States remained protected from aerial bombardment by the ocean, by the inability of planes to fly across it without refueling. Even so, in 1941 New York began to prepare for a blackout, too. General civil defense planning had been under way for months. "There is said," reported the *New York Times*, "to be a genuine fear that the American people would embark on innumerable programs for the defense of their home cities — programs that might have no merit beyond the enthusiasm of the promoters. The [commission set up by the War Department aims] to forestall, by planning, the bungling of improvizations [*sic*]."

Manhattan was already subdued by the war, by rationing, and by the dimming of the lights in Times Square and along the waterfront, ordered to prevent the silhouetting of vessels in the docks and shipping lanes, which would have made them easier targets for German U-boats. Still, New York was the premier electric city, and though officials had the example of London to emulate in its methods for blacking out, it took more than a year, and numerous district drills, to prepare the borough of Manhattan — with its hundreds of miles of streets, its fourteen thousand acres — for its first complete drill. The day and time of the beginning of the drill — 9:30 P.M. on May 22, 1942 — and its duration — twenty minutes — were widely advertised beforehand and entirely expected. It turned out to be a foggy night with a cool wind. In Times Square, immediately before the air raid drill, wardens and policemen began to call, "Get off the streets. . . . Everybody off the street." Pedestrians crowded into doorways and under marquees. They piled into the subway entrances. According to the *New York Times*:

> The crowds melted into the darkness, taking refuge wherever possible. You could stand in the center of Broadway or Seventh Avenue and barely discern a moving form. Several rooms in the Claridge Hotel still showed pale light. The wardens and policemen shrilled their whistles or called hoarsely to these places. "Lights out, Claridge! Put out those lights!" The cry went up from male and female voices, and one by one the Claridge lights died. . . . Now and then a man or woman, or a couple, broke for cover and the patter of their feet was clear and distinct above the whispers of those already under shelter. . . . Kiosk lights at Forty-first Street and Seventh Avenue were blurry globes in the rain. A zealous warden, unable to find some one to put it out, dumped a basket of rubbish over one of these lamps but the beams lanced out from under the blanket and threw shafts into the shining gutter.

There were no lights all the way up Fifth Avenue, throughout Greenwich Village, in the fog and rain of Harlem, or along the crooked, narrow streets of Chinatown. On the East Side, shades were drawn over the Sabbath candles in the windows. Millions breathed in the dark, sitting in living rooms or standing at the sink or in entry halls, on the dance floor, or by their work stations. Although there was no order for quiet, few spoke above a whisper.

When the all clear was sounded at 9:50 P.M. and the lights came on, they hadn't been doused long enough to give people's eyes time to adjust to the dark. Chemical changes in the retina had not yet occurred. In Times Square, almost immediately after the drill, voices rose above the noise of traffic starting up, and dance music leaked out of nightclubs. Crowds poured out of the entryways and up the subway stairs and moved steadily along the streets once more. "As the lights came on again in hotels and shop windows and traffic lamps winked red and green through rain, the crowd cheered."

In London, when after nearly six years of nighttime restrictions the blackout order was lifted, the exhausted populace didn't seem to have the heart to light up the city again. The *New York Times* reported, "For every undraped window there were twenty in darkness. Some of the black-out windows showed chinks of light, which hitherto would have brought the air raid warden to the door. . . . Department stores stayed dark, as did the electric signs on Piccadilly. It will be some time before wiring can be refurbished." The streetlamps also remained dark, because their wires needed to be refurbished as well. "The few householders who left their windows bare had to remember that a front room from the street looked like a well-lighted stage set and act accordingly. Mostly they blacked out, as before."

16

LASCAUX DISCOVERED

DURING THE LONG MONTHS in which the lights were out all over Europe, the Paleolithic drawings in the Lascaux Cave were discovered. On September 8, 1940, in the Vézère Valley in the Black Périgord of France — then known as Vichy France — seventeen-year-old Marcel Ravidat, along with several friends and Ravidat's dog, were roaming the hills above their town. In the nineteenth century, the land had been cultivated with grapevines, but when the vines were killed by phylloxera, local farmers dug them up and planted the area in pines. When one of the trees toppled over early in the twentieth century, it revealed an opening in the ground about the size of the entrance to a fox's den, which the farmers blocked off to protect their cattle from possible injury. Legend has it that during Ravidat's walk, his dog fell into the hole, and when Ravidat scrambled to the dog's rescue, he noticed an opening to a deep shaft. He and some friends returned four days later. "I made myself a very rustic but quite adequate lamp from an old oil pump and a few meters of string," Ravidat recalled. "When we arrived at the hole I rolled some large stones into it and was surprised at the time they took to reach the bottom. . . . I set to work with my big knife . . . to widen the hole so that we could get into it." After hours of digging, planning, and crawling, they arrived at the floor of the cave. "We raised the lamp to the height of the walls and saw in its flickering light several lines in various colours. Intrigued by these coloured lines, we set about meticulously exploring the walls and, to our great surprise, discovered several fair-sized animal figures there. . . . Encouraged by this success we began to go through the cave, moving from one discovery to the next. Our joy was indescribable."

In the following days, other boys came to explore the cave, as did the local schoolmaster and then local men and women. Within a few weeks, people from all over the region began to arrive — more than five hundred visitors in the span of one week. "Like a trail of gunpowder the rumor of our discovery had spread through the region," Ravidat said. Old women brought their own candles to see by. They walked over rough ground and climbed down the narrow entrance. The paintings then, seen by the light of rudimentary lamps and small open flames, must have appeared much as they had in the Pleistocene.

Scientists and archaeologists came as well and mapped the cave: Chamber, Hall, Gallery, Passageway, Apse, Shaft, and Nave. They named the paintings: Frieze of the Black Horses, Frieze of the Small Stags, Procession of Engraved Horses, Frieze of the Swimming Stags, Niche of the Felines. After the war, the number of visitors to Lascaux increased markedly, and in time a walkway was put in.

During the thousands of years that the Lascaux Cave had remained undiscovered, the temperature inside never rose above 59 degrees, and the humidity level stayed constant. When the cave was crowded with visitors, the temperature sometimes rose to nearly 90 degrees. In 1955 excess carbon dioxide, produced by the visitors' breath, caused the first noticeable signs of deterioration in the paintings. Water droplets began to appear on the walls, and as they trickled down, they erased the pigments on the backs and necks of the animals. In 1958, to mitigate this problem, an air-exchange machine was put in, but it also worked to scatter the pollen that came into the cave on the visitors' feet. As a result, algae — "green leprosy" it was called — began to ravage the paintings. The animals were disappearing "in a prairie" of algae, Ravidat recalled. Also apparent was the "white disease" — crystals of calcite encouraged by the increased levels of carbon dioxide, humidity, and temperature — which began to cloud the paintings. To protect them, the cave was closed to the public in 1963.

In 1981 Mario Ruspoli was asked by the French Ministry of Culture to make a cinematographic record of the Lascaux paintings. It took him years to complete his work, since he was allowed access to the cave for only twenty days a year, in March and April, when the cave was at its coldest. His crew could work for only two or three hours at a time so that the heat emanating from their bodies and their hand-held, 100-watt quartz lamps could dissipate. Just two of their lamps could raise the temperature by several degrees and also raise carbon dioxide and moisture levels. One human body gave off more heat than the lamps. Ruspoli recalled:

> The lights were never held on a particular spot for longer than twenty seconds, and at the end of each take they were turned up to the ceiling or down to the floor, causing the image to fade into darkness. . . . After shooting it was advisable not to light

the lamps for a little while in order to allow the slight rise in temperature caused by the bodies and the quartz lamps . . . to disperse.

Our precision lenses sometimes surpassed the powers of perception of the naked eye, bringing out details which were only just discernible, particularly in the painted surfaces and around the figures. . . . At first it seemed that it would be impossible to film with so little light . . . but in actual fact the opposite proved to be true. . . . Our modest resources and the restricted lighting that was permitted made us take a new cinematic approach to the art on the cave walls. . . . We had to use swift, precise and spontaneous takes, the camera moving forward through the dark cave and disclosing the space as it emerged. . . . This slow unfolding of the images in the silence of the cave took us to the edge of another world . . . and we ourselves gradually began to feel like initiates. . . . The Upside-down Horse curves round a pier and the Great Black Auroch makes use of the curious relief of its concave niche: when it is seen at an angle from the end of the Gallery, only its head is visible; the body is concealed behind a projection in the rock and is only revealed when one moves towards it. . . . We noticed all this as we advanced, lamps in hand, along the wall toward the back of the cave. The painted figures emerged gradually from their hiding-places in the rock and this movement made them seem to come alive. . . . To the members of my team and myself, Lascaux became a sort of second homeland.

PART IV

Science tells us, by the way, that the Earth would not merely fall apart, but vanish like a ghost, if Electricity were suddenly removed from the world.

—VLADIMIR NABOKOV,
Pale Fire

Nothing, storm or flood, must get in the way of our need for light and ever more and brighter light.

—RALPH ELLISON,
Invisible Man

17

BLACKOUT, 1965

> ... we have built the great cities; now
> There is no escape.
>
> — ROBINSON JEFFERS,
> "The Purse-Seine"

T HE RURAL ELECTRIFICATION PROGRAM slowed to a
near halt when supplies and manpower were redirected
to fighting World War II, but once hostilities ceased,
the electrification of the American countryside resumed. By
1960, on the twenty-fifth anniversary of the Rural Electric Administration, 96 percent of American farms were connected
to electric lines. The average rural customer used about 400
kilowatt-hours of electricity per month, compared with the
1935 average of 60 to 90 kilowatt-hours per month. Although
farms continued to disappear, rural lines connected more and
more people as potato and beet fields, pastures, apple orchards,
and orange groves were plowed under and turned into suburban neighborhoods. With the advent of nuclear energy, it was
rumored that electricity would become too cheap to meter.

During these postwar years, the U.S. power industry remained stable. The New Deal regulations were still in place,

and the industry grew at a steady 7 to 8 percent per year. The utility companies had come to be thought of as natural monopolies, and the power grid had reached a size and significance that could never have been imagined in the late nineteenth century, when a writer at *Harper's*, commenting on the accomplishment at Niagara, declared, "It is scarcely to be expected that current can be brought as far as New York [City] to commercial advantage." Individual power stations, including Niagara, had now grown to serve areas that could encompass entire states, and not one of them stood alone: each was connected to, and might borrow from, a host of others in an electric grid. The point of generation was often far from the point of demand, and the interlacing corridors of long-distance wires that cut through rough, quiet, country linked farms, cities, and suburbs — places with ever-changing historical realities and interrelationships — in a shared fate.

In each power station, watchers in the command center hunched day and night over consoles, scanning screens, dials, and gauges that monitored turbines and took account of current running back and forth over thousands of miles of lines. Such a network proved economical: "In times of normal demand for electricity the member companies [could] shut down some of their expensive steam-fed facilities and 'ride' on the cheaper current provided by hydroelectric generators." And on the whole, it was more reliable. If, for instance, a generator had to be shut down for maintenance or repair at an eastern Massachusetts station, power could be borrowed from New York or Wisconsin, for there was virtually no electrical distance between them. Power could be generated in New York and travel to New Jersey before arriving in Massachusetts if need be. While all this was being accomplished, people in Boston might notice nothing more than a momentary flicker of their lamps.

But for all the massive reach of electricity, the generation of electric power also stood in the same delicate balance as in 1910, when Edward Hungerford detailed the way a cloud could stress the power systems of New York City, for the need to maintain an equilibrium of supply and demand hadn't changed. There was more at stake, of course. The balance had to be maintained across numerous power stations, and — since electricity moved back and forth across the wires — surges, flow reversals, or disruptions at one power plant could have far-reaching ramifications and might ultimately affect the synchronicity of the entire system.

That synchronicity was essential. By 1965 all public utility generators east of the Rocky Mountains ran in sync with one another so that alternating current could be seamlessly switched from one generator to another throughout the system. You might think of their working sound as the music of our spheres, for if even one were to fall out of phase and begin spinning at its own speed, if its steady, precise humming became discordant — a wobbly song of its own — well, then . . .

A slight variation [could] be tolerated if it [was] soon brought into line. A major variation [would force] other generators to "hunt" for a new phase more aligned to the maverick's. . . . The out-of-phase current finally [would cause] other generators on the circuit to shut down. The more generators that cut off, the more that [would] follow suit. For any generator feeding current into the system at that point would be so overloaded that its safety devices, the circuit breakers, would bring it to a halt.

What goes on across the power grid, it's said, "is like a game of tug of war, which works as long as neither side — the generating stations and the load centers — wins. If one side falters, and the rope moves too far, everyone on the other side will fall down."

All through the brief daylight hours of November 9, 1965, the forty-two interconnected power stations between Ontario and Boston that made up the Canadian and U.S. Eastern Interconnection hummed along. There were no extraordinary demands on the supply: the weather was mild and the sky clear. As the sun set shortly before five o'clock, farmers in the countryside, with their fields all plowed under and their barns full of hay, were beginning the evening milking. In small towns, stores flipped over their Open signs and closed up shop. Everywhere, wives and mothers began preparing dinner while children sat transfixed in front of the TV watching *The Three Stooges*. City office workers, their day done, jammed the elevators, subways, escalators, streets, and trains. Car lights formed brilliant rivers down avenues and across bridges, their drivers obeying, anticipating, or trying to beat the red, amber, and green signals that had been directing the flow of traffic ever since the first four-way, three-color stoplights — based on controls used by railroads — were devised in the 1920s.

Nowhere were more people in transit than in New York City, when, at 5:16 P.M., more than three hundred miles to the north, at the Sir Adam Beck No. 2 generating station in Ontario, a relay — a device about the size of a telephone of the time which automatically regulated and directed the flow of current — failed to give off the proper signal. As a result, a circuit breaker did not open, which caused excess electric current to surge through the system. According to John Wilford and Richard Shepard,

Because the relay did not work, there was an overload on the line. This caused relays on other lines feeding through the plant to operate circuit breakers and the total of 1.6 million kilowatts going through the Beck station suddenly reversed course — as electricity will do when it is unable to flow in the direction it is supposed to.

Much of all this vast quantity of electric current raced back across upper New York State, tripping safety equipment from Rochester to Boston and points beyond. At this point the second phase in the breakdown occurred. Consolidated Edison in New York City, and other power companies to the south that had been receiving power from the area knocked out of service by the power surge, were hit by a reverse flow in their own lines. Their power rushed, somewhat as air will rush to fill a vacuum, into the upstate New York–New England–Ontario region. The generators in New York City and elsewhere, inadequate to fill the huge power vacuum, automatically shut themselves off.

Twenty-eight of the forty-two power plants in the region shut down, and the darkness sped south and east within the span of just over twenty minutes. At 5:17 Rochester and Binghamton, New York, shut down. Then eastern Massachusetts, the Hudson Valley, New York City, and Long Island lost power. All of Connecticut shut down at 5:30. Parts of Vermont and southwestern New Hampshire — the last to go — went dark at 5:38. The plant on Staten Island maintained power because it was able to break free of its network connection before failure. This good fortune was as bewildering as the bad:

In the New York State system . . . the 345,000 volt lines . . . are so designed so that a region hit by a local power failure can immediately have a surge of energy sent to it from . . . another power source. . . . To be capable of this instantaneous action, the system must be able to accept a wide range of power loads. Hence the main trunk, the electric superhighway bisecting the state, is not equipped with circuit-breakers sensitive to slight changes in load. The decision to cut a local system out of the grid is a human one and the actual cut-off must be done manually at a local control center.

But on Staten Island, for some reason, a circuit breaker tripped unexpectedly, automatically severing it from the rest of the grid. The manager of system operations there could only say, "I don't know why it opened."

Northern New Jersey, Pennsylvania, and Maryland had systems that carried lesser voltage, and their circuit breakers were set to trip automatically, which severed them from the grid in time. The state of Maine's power system was only weakly connected to the rest of New England's and so was able to cut itself off from the failure and subsequently lend power to parts of New Hampshire. The lights in these areas formed a fringe of illumination around a vast darkness: almost everyone across 80,000 square miles of the northeastern United States and part of Ontario — 30 million people — had lost their electricity.

At the time of the breakdown, there was no apparent reason for the outage — no storm, no high winds or lightning, no trees touching high-tension wires — and the cause would not be known for days. In each power plant, engineers and technicians were left to wonder whether something in their own system had triggered the shutdown, while people in the countryside — accustomed to occasional local outages even in good weather — naturally thought that maybe a car had hit a pole somewhere down the road. In the cities, there were vague notions of sabotage: "'The Chinese,' a housewife on New York's East Side thought when she saw New York fade from her window, and then was a little ashamed." And "through the minds of two knowledgeable newspapermen flashed the same thought at about the same time, as they were to discover later.

Both thought, 'The anti-Vietnam demonstrators have pulled something off.'" Some said it was an earthquake; others recalled extraordinary times. "I could see the New York skyline from my windows," remarked a woman from Brooklyn. "All of a sudden, it's dark — dead, kind of. The last time was in the war, it was dark about the same way."

In New York City — the world's most concentrated electric market — 800,000 people were trapped in the subway; countless others were in elevators — "like hamsters in their cages," a *New York Times* reporter would say — or in offices high up in skyscrapers. Those riding on escalators "glided down more and more slowly, until, at last, they were scarcely moving at all." Not everyone risked descending to the street by way of the darkened stairwells. More than five hundred people would spend the night in the forty-eight-story skyscraper that housed the offices of *Life* magazine, and an emergency medical center would be set up in the lobby.

Those already in their cars and on their way home had limited fuel, since gas pumps needed electricity to run. All the stoplights failed, and although some citizens tried to direct traffic and policemen set flares in the roads at dangerous forks and intersections to help drivers negotiate their way, most of the city was quickly snarled in gridlock. Some native New Yorkers walked across the bridges — flashlights and transistors in hand — for the first time in their lives. Others caught rides by hooking onto the back bumpers of crowded buses. Cabbies hiked up their fares. A. M. Rosenthal wrote, "As usual New Yorkers helped gouge themselves. They stood in the roadway, flagged down taxis and shouted 'Thirty dollars to Brooklyn!' 'Ten dollars to the Village!'" It would be said of that night that it was easier to cross the Atlantic to Cairo than to get to Stamford, Connecticut, from the city.

* * *

"The more efficient the technology, the more catastrophic its destruction when it collapses," observes Wolfgang Schivelbush. This was a given, and although utility executives and engineers always acknowledged that a widespread failure of the grid could occur, few believed that it would, and they'd made no contingency plans for an extensive, cascading failure. Their confidence had fostered a sense of complacency: out of 150 hospitals in New York City, fewer than half had adequate backup power. Doctors had to perform emergency surgeries by flashlight, and five babies were born by candlelight at St. Francis Hospital.

Likewise, the airports were entirely unprepared for the loss of power. They had no radar for six hours and no field lighting. High above the city, airplanes lost their ground orientation and were unable to land. "It was a beautiful night," recalled one pilot. "You could see a million miles. You could see the Verrazano Bridge and parts of Brooklyn, but beyond Brooklyn, where we usually see the runways at Kennedy and Floyd Bennett Field it was dark. . . . I thought 'another Pearl Harbor.'" Kennedy International Airport closed down for almost twelve hours, though several hours into the blackout, LaGuardia was able to light one runway with power from a water-pump generator. Both New York airports had to cancel or divert about 250 flights; some had to be rerouted as far away as Bermuda.

The fine, clear voices that ordinarily gave the news of world — the dead in Vietnam, the protesters at home, the condition of former president Dwight D. Eisenhower's heart — turned tinny and staticky, reduced to the sound on transistor radios. The first reports that came through were wildly inaccurate, claiming that the blackout stretched all the way to Miami, that it reached to Chicago, that Canada lay in darkness. The fears would not be allayed for several hours. "We still knew nothing about what had really happened, what had created our predicament," recalled a *New Yorker* writer, "but just then anybody who might still have been worried that the blackout heralded a foreign takeover was reassured by an announcement from the Pentagon, over the little transistors, that it had no effect on our 'defense posture.' . . . The power companies soon provided similar comfort, with their talk of 'outage.'" Still, rumors would live long after the lights returned.

The true quiet of the world felt strange, "as if the darkness had somehow smudged away the horns and the other noises of the traffic." Electrical sounds, like Pythagoras's music of the spheres, had always been in people's ears and were what they took for silence. In the relative hush, suddenly a million little things were in danger of perishing. Damp glass greenhouses began to cool down. At the Bronx and Central Park zoos, "the men, working without sleep, stuffed blankets between the bars in the small-mammals house, where diminutive, heat-sensitive lemurs, flying squirrels, and small monkeys began their nocturnal peregrinations. The reptile house presented a difficult problem, since no one was willing to try to wrap a cobra in a blanket. Small portable propane gas heaters were taken in to warm the cold-blooded vipers, anacondas, iguanas, caymans, crocodiles and their ilk." It may have been too cold for iguanas, but the temperature outside was perfect for storing blood — between 38 and 41 degrees — so hospitals and blood banks took their supplies to the roofs for keeping.

Night was truly night again, just as in the Middle Ages, and, also as in the Middle Ages, light became precious once more. People struck match after match to light their way down flights of stairs: "Two matches, carefully tended, were enough to light the distance between one floor and the next. Walking down eighteen flights to the lobby, we used exactly thirty-six matches." People shared candles with one another, and gougers sold them on the streets. Tapers stuck in beer and wine bottles or tea lights set on saucers illuminated cold meals in homes, restaurants, and coffee shops, as well as a banquet in the Astor Ballroom. They flickered alongside pool games and across the faces of actors preparing for a performance — the lights, after all, could come back on at any time. They burned in newsrooms and at newsstands, in firehouses and police stations, on the mayor's desk and beside card games on trains. Wax dripped onto tabletops and onto floors; days later newspapers would publish instructions on how to remove it from surfaces.

Just as the lights went out, the moon, a day after full, was rising:

> The moonlight lay on the streets like thick snow, and we had a curious, persistent feeling that we were leaving footprints in it. Something was odd about buildings and corners in this beautiful light. The city presented a tilted aspect, and even our fellow-pedestrians, chattering with implacable cheerfulness, appeared foreshortened as they passed; they made us think of people running downhill. It was a block more before we understood: The shadows, for once, all fell in the same direction — away from the easterly, all-illuminating moon. . . . We were in a night forest, and, for a change, home lay not merely uptown but north.

Without that moon, the night of November 9, 1965, would have been very different. Air tragedy, it was said, had been averted because its light, along with that made possible by auxiliary power in the main control towers at the airports, was enough for pilots already in descent to see by. The previous night, rainstorms had soaked the region, and clouds had covered the moon and stars. Had the lights gone out then, there surely would have been more than one disaster. As it was, emergency rooms filled with people who'd been hit by cars or tripped on the sidewalks. There were pockets of looting, but by dawn there would be less crime reported than on an ordinary November night.

Time and task were both disorienting, for if you were to remove everything from our lives that depends on electricity to function, homes and offices would become no more than the chambers and passages of limestone caves — simple shelter from wind and rain, far less useful than the first homes at Plymouth Plantation or a wigwam. No way to keep out cold, or heat, for long. No way to preserve food, or to cook it. The things that define us, quiet as rock outcrops — the dumb screens and dials, the senseless clicks of on/off switches — without their purpose, they lose the measure of their beauty, and we are left alone in the dark with countless useless things. Skyscrapers take on a geological sheen, and the stars resemble those of ancient times.

Yet unlike in ancient times, people weren't accustomed to giving in to the long November night. For most, the dark wasn't restful; it simply felt as if the world had stopped and everyone and everything were suspended in amber, especially after the novelty of the first hour wore off. For as long as no one had any idea at all how long the helplessness would go on, there was no future, and no knowing the future. After a few hours, theaters canceled their scheduled performances, and people ran out of pocket money. They were still lined up outside phone booths waiting to call home, but what could they say other than that they were somewhere? November 9–10,

1965, became known as "the night of the long night," and it was particularly long for those trying to sleep in hotel lobbies or on office floors; in barber chairs or on cots in banquet rooms; curled up in hallways or sprawled on subway stairs or benches in train stations.

Meanwhile, throughout the affected area, each local utility had become an island again, and in each affected power station not only were the managers looking at systems that had no obvious failure, but they were also still unsure as to whether their own station was at fault or merely one link in the cascade. They had to get their system back up with the same equipment that had shut it down, and although it took only a few seconds to lose power, it would take hours to get back online — for it's no simple thing to align the spheres again. All switches, relays, and circuit breakers had to be checked, as did turbines, generators, and boilers. "The turbine generators had to be turned slowly by mechanical means to make sure they had not been bowed out of shape in the blackout." The power failure itself had caused some damage. For instance, turbine bearings at Con Edison's Ravenswood plant were damaged by lack of lubrication during the lapse in power.

Power was needed to beget power. "Unfortunately many of the affected utilities had made no provision for the unlikely possibility that their entire system would shut down simultaneously and, hence, there were no independent auxiliary power sources for such an eventuality. Intricate circuits had to be established, some from remote sources, to feed in the essential auxiliary power." Even with power, the enormous boilers, some of which were as tall as fifteen-story buildings, had to be heated up to 3,000 degrees, and the pressure had to be built up to more than 2,000 pounds per square inch. And everything couldn't be turned on at once, or it would overload the system. "As power became available, it was essential that the load be picked up in a careful, sectionalized, synchronized process. As each section was brought up to load, it was necessary to synchronize its frequency with that of the energized remainder of the system. It was then possible to tie the section in with the remainder of the network without disturbing the maintenance of the network's synchronism."

Service was restored to parts of New York and New England within a few hours, but it would be almost midnight before northern New York State was completely back online; Boston and Long Island didn't get power until 1:00 A.M. In New York City, it would take more than thirteen hours for the power to fully return.

The electric lights of New York — the gaudy marquees and overlit skyscrapers, which had for decades far exceeded necessity — accounted for a small fraction of the overall power demand. Even so, it was by light that most people had come to gauge their connection to life, and it was the loss of light that was most remarked upon. The following day, the headline of an Italian newspaper read: "New York Cancelled by Darkness."

Sometime after three in the morning, as in section after section of the city signs of a world coming to life again registered in little whirrings and tickings, faint and then full, the editor of *Life* magazine noted: "Ralph Morse, who had taken the first pictures of the blinded city from a 28th-floor window, now began to take the last pictures from the same position. Slowly, during the next 1½ hours, the city came alive again, a blaze of lights here, a blaze there. . . . Morse's camera caught the radiant rebirth."

* * *

That morning, subway workers had to comb all 720 miles of track before the trains could run again, just to make sure no one had fallen and lay injured on the rails or was lost and wandering along the lines. Gas crews went from house to house to check the pilot lights — which were powered by electricity — in the stoves and boilers of every customer. The current of weary people who'd spent the night in the train stations flowed past people coming to work again.

Perhaps it had been hardest on the old and the sick, who'd had a nerve-wracking time. For a few, the dark was fatal: one man was found at the bottom of an elevator shaft, still clutching the nub of a doused candle. For others, it was a night unlike any other in its generous and quiet beauty. Among those who spent hours playing cards and drinking whiskey or making small talk in dark offices and subway cars with people they sometimes couldn't even see, some struck up a camaraderie they would never have had by any other light. "Everybody recognizes everybody else now," one woman said. "Although they've seen me for ten years and they've done nothing but help me up the stairs, now it's a tip of the hat and a 'good morning, Phyllis, how are you today?'"

The lost hours eventually faded into a strange dream full of quirky things, though there were moments that would be intensely remembered afterward — of lighting grease pencils to see by, of being given coffee and pastries by transit workers while waiting in a darkened subway car, of the sheen of the moonlight on the side of a skyscraper.

The blackout of 1965 spurred the first serious examination of the electric grid and its fragility. The subsequent Federal Power Commission report, besides advocating extensive changes to the grid system itself — ones that were hoped would both strengthen the grid and confine future outages — recommended backup energy supplies for airports, hospitals, elevators, gas stations, and radio and television stations; auxiliary lighting for stairways, exits, subway stations, and tunnels; subway evacuation and traffic control plans. But even with such measures in place, in July 1977, when a series of lightning strikes sent an enormous surge through New York City's power system, circuit breakers — which were designed to reset automatically — failed to close, and the city was plunged into darkness again.

Although in many ways this outage was similar to the one twelve years before — the stalled subways and traffic, the gouging, the camaraderie among people stuck together, the kindnesses (restaurants set up tables on the sidewalk and stayed open; a bagpiper played in Grand Central Station) — the city was a different place, the age a different age. Unemployment among young men in some of the black and Hispanic neighborhoods exceeded 40 percent. The night was hot and muggy — sweltering — and the pale sliver of a new moon set before the lights went out at 9:34 P.M., so there was no consoling light reflecting off the skyscrapers, nothing but the torch on the Statue of Liberty to relieve the blackness. Looting broke out in all boroughs of the city, and arsonists set more than a thousand blazes. After a few hours, thieves even began stealing from the looters in a free-for-all that continued beyond the twenty-five hours it took to restore power. Police arrested thousands, and hospitals were swamped with people cut by knives and glass. Three people died in the fires, and a looter was shot dead. In a number of hospitals, the emergency backup systems, which had been made mandatory after the 1965 blackout, failed to work. Doctors stitched wounds by flashlight, and nurses resorted to squeezing air bags by hand for patients who were dependent on respirators.

In writing of it later, no one waxed poetic about the moon

on the buildings. It was the looting that dominated people's thoughts. "We are in much worse trouble than we thought," commented a writer for *The New Yorker.* "In the blindness of that night, New York and America could see rage. We've been put on notice again. We may continue to ignore the terrible problems of poverty and race, but we must do so aware of the risks to both justice and peace."

The size of the machine that had become us had grown to be almost incalculable — some would say it was the largest machine in the world. Yet when it failed, societies were pervaded by the same feeling that those who had experienced the loss of gaslight in the nineteenth century had: of being vulnerable, of having given over control of our life. Russell Baker, writing in the *New York Times* after the 1965 blackout, imagined the ultimate fragility of the electric grid:

> The end came on Sept. 17, 1973. It had been forecast by an M.I.T. undergraduate who had been running the law of probability through his computer. . . . The chain of events on that last day began at Shea Stadium at 4:43 P.M. when the Mets finished a scoreless ninth inning against the Mexico City Braves, thus becoming the first team in history to lose 155 games in a single baseball season. . . . Two minutes later, Irma Amstadt, a Bronx housewife, turned on the kitchen faucet and noticed that there was no water. Going to the telephone, she dialed her plumber, not knowing that at that very moment, in defiance of probability, 6,732,548 other persons in New York were simultaneously dialing telephone numbers. . . . Mrs. Amstadt's call was the one that broke the system's back.

The grid can be as fragile as Baker imagined it to be. In August 2003, during hot weather and high demand, transmission lines — as transmission lines will do when they heat up —

expanded and sagged all over the grid. In Walton Hills, Ohio, sagging wires touched some overgrown trees beneath them, which began a chain of events that plunged 50 million people in the eastern United States into the dark. It was the largest blackout in American history.

IMAGINING THE NEXT GRID

BY 1965, THE SAME YEAR as the Northeast blackout, New York artist Dan Flavin had turned to fluorescent light as the sole medium for his work. "Regard the light and you are fascinated — inhibited from grasping its limits at each end," he wrote in December of that year.

> While the tube itself has an actual length of eight feet, its shadow, cast by the supporting pan, has none but an illusion dissolving at its ends. This waning shadow cannot really be measured without resisting its visual effect and breaking the poetry.... Realizing this, I knew that the actual space of a room could be broken down and played with by planting illusions of real light (electric light) at crucial junctures in the room's composition. For example, if you press an eight foot fluorescent lamp into the vertical climb of a corner, you can destroy that corner by glare and double shadow. A piece of wall can be visually disintegrated from the whole into a separate triangle by plunging a diagonal of light from edge to edge on the wall.

For the next thirty years, Flavin used standard fluorescents in the available colors of blue, green, pink, red, yellow, and four kinds of white to explore everything about light save for its utility: the interplay of light and space, light and solids; the way colors mingled; the way glare and shadows dispersed solidity. He understood light as an endless and intricate medium for his work, yet he also knew that without the stability of infinite electrical connections, his works — like our own ordinary lights — were no more than heavy, inanimate objects made of glass and metal. "Permanence just defies everything," he once said. "There's no such thing. . . . I would rather see [my work] all disappear into the wind. Take it all away. . . . It's electric current with a switch — dubious. . . . And rust and broken glass."

Throughout the decades of Dan Flavin's career, the connections essential to his work became more and more dubious, and not only because of power outages. By 1973 the economies of the United States, Europe, and Japan relied on abundant, cheap oil. It fueled an insatiable car culture, yes, but oil was also essential to the energy grids of industrialized countries. For instance, it accounted for 20 percent of the fuel used for electricity generation in the United States. "Oil had become the lifeblood of the world's industrial economies," Daniel Yergin observed, "and it was being pumped and circulated with very little to spare. Never before in the entire postwar period had the supply-demand equation been so tight." Not only was there little to spare, but a good share of the oil consumed by the West and Japan was imported from the Middle East, and when Saudi Arabia instigated an oil embargo in the fall of 1973 in response to American arms shipments to Israel, fuel supplies tightened throughout the world. By December the price of oil, which sold on the world market for under $6 a barrel in early October, had nearly tripled in price.

Suddenly, as the country headed into winter, the "liberty poles" so valued by American farmers, and which seemed so strong and enduring in the landscape, proved to be entirely vulnerable. To conserve existing fuel supplies, President Richard Nixon, in addition to calling for restrictions on heating fuel and gasoline and setting lower speed limits on the interstates, called for the conservation of electricity. Specifically, Nixon called for the dimming of nonessential lighting such as advertisements and all decorative Christmas lights, both public and private, including the lights in New York City's Times Square. Although the electricity required for decorative lighting accounted for only 2 to 3 percent of all energy consumption in New York City, and lighting in general was responsible for about 6 percent of energy use nationwide, officials hoped that the dimming of such lights would encourage citizens to conserve energy in their own homes. "It's very sad to be a party to darkening a city so renowned for its lights — it's just heartbreaking," the municipal service administrator for New York City commented. "But it has a psychological effect, because it's difficult to get someone to turn down his thermostat if he sees lights blazing in a public place."

Although the shared and the celebratory — the flamboyant lights of advertising and the seasonal lights of the holidays — may seem to be the most dispensable of things in practical times, they draw our eyes out of the sea of lights we live in, and they take on outsize significance when doused, as if something essential has been taken away from the culture, especially in winter, when artificial light has always had its greatest meaning. And in 1973 their dimming *did* mean that something essential had been taken away, something larger than sheer illumination: the assumption that we could live without thinking about energy, that we could take it all for granted.

Writer Jonathan Schell understood that the damped-down

world of the oil embargo was also a world whose underpin-
nings had profoundly changed: "This winter as the nation sits
in its dimmed, chilled, living rooms watching the comet Ko-
houteck, which is due to appear in our heavens soon (it will be
our finest Christmas ornament this darkened season) . . . the
newly recognized global limits of natural resources . . . force
us and the Arabs and Europeans and the Japanese and all the
rest of the peoples on the planet into dependence on one another. In the last analysis, the rationing we need is global."

And yet if the solution seemed to be global, it could also
be personal. Who could blame some for wanting out, for following Helen and Scott Nearing back to the land and into
"the Good Life," as did thousands of city dwellers in the late
1960s and early 1970s? The back-to-the-land movement was a
response not only to the energy crisis of the time but also to
the growing separateness from nature of modern life, with its
inevitable interconnections. Poet Baron Wormser, along with
his family, lived "off the grid" in the rural Maine woods for
more than twenty years. As Wormser experienced it, the light
of his kerosene lamps belonged to a different time, a different
kind of evening. "Night's coming was so profound, so transfix-
ing, so soft yet indelible that I was startled and lulled in the
same awed moment," he wrote. "I remember very clearly feeling how, second by tiny second it was getting dark, how the
dark was creeping in, how it was inexorable and delicate."

He acknowledged that he romanticized the same lamp that
had been so eagerly buried by rural folks in the 1930s:

A few guests over the years found the stench appalling and
the light feeble. As much as they wanted to be charmed, they
weren't. I loved lying in bed and reading by the light of a
small kerosene lamp. I was reading in the presence of an actual
flame. . . . Time was steady, but in the flame's movements it var-

ied. . . . It was a romantic glow. . . . The trembling light is quietly breathtaking. It causes soot and stench; it came from the
hard work of mining, processing and trucking. . . . All true, but
the feeling remains. Touch the glass chimney — it is hot with
the heat that signals light.

However much he romanticized lamplight, Wormser also
came to understand some of its costs and effort, which modern society at large would have to acknowledge sooner or later.
"Light did not materialize itself," he wrote. "Our efforts each
day made it happen. A match had to be struck. Our heedless-
ness had a limit."

Although Middle Eastern oil producers eventually lifted their
embargo in March 1974, after Israel agreed to pull its troops
out of the Sinai Peninsula, the price of oil remained higher
than before the embargo, and the stability of fuel supplies fluc-
tuated for years afterward. When President Jimmy Carter
took office in 1977, he made energy independence a major
goal of his administration. Carter, in a time before the widespread recognition of the effects of fossil fuels on climate,
imagined exploiting the known coal reserves of the United
States in order to alleviate the country's dependence on foreign oil. He also stressed conservation, appearing on television
in a cardigan sweater and urging people to turn their thermo-
stats down to 55 degrees at night, and he planned legislation
that would foster the development of cleaner, more efficient
energy generation.

"We simply must balance our demand for energy with our
rapidly shrinking resources," insisted Carter in an address to
the nation in April 1977.

By acting now we can control our future instead of letting the
future control us. . . . Our decision about energy will test the

character of the American people and the ability of the President and Congress to govern. This difficult effort will be the "moral equivalent of war" — except that we will be uniting our efforts to build and not destroy. . . . The 1973 gasoline lines are gone, and our homes are warm again. But our energy problem is worse tonight than it was in 1973 or a few weeks ago in the dead of winter. It is worse because more waste has occurred, and more time has passed by without our planning for the future. And it will get worse every day until we act. . . . The world has not prepared for the future. During the 1950s, people used twice as much oil as during the 1940s. During the 1960s, we used twice as much as during the 1950s. And in each of those decades, more oil was consumed than in all of mankind's previous history.

As part of this effort, Carter signed the National Energy Act into law, a component of which, the Public Utility Regulatory Policies Act (PURPA), marked the first significant legislation to affect the power grid since the New Deal. PURPA permitted independent power producers that met strict fuel efficiency standards to enter the electricity market — previously the sole domain of the utility companies. Carter hoped that such legislation — and the competition it might engender — would encourage the construction of more efficient coal plants and the development of alternative energy sources such as wind, solar, and biodiesel. PURPA didn't succeed in fostering the significant development of alternative energy sources, but it did open the way for more deregulation of the energy industry.

The Energy Policy Act of 1992, signed by President George H. W. Bush, encouraged further deregulation of the electricity industry at the state and federal levels and broadened the types of competitive companies allowed to produce energy for the grid. It brought energy trading companies such as Enron, Dynegy, and Reliant Energy into the mix. These companies were essentially unregulated power brokers that might not own any generating facilities at all. They bought and sold electricity on the open market and also traded in derivatives, betting on supply and demand as with any other commodity, such as corn or pigs. Enron CEO Jeffrey Skilling claimed the development would be a boon to consumers: "We're working to create open, competitive, fair markets," he remarked. "And in open, competitive, fair markets, prices are lower and customers get better service. We are the good guys — we are on the side of angels."

Decisions concerning the deregulation of power grids rested largely with state governments, and during the late 1990s — a time of fervent support for deregulation in general — a number of states, including California, passed legislation that deregulated their power industries. California's electric rates were historically high, and its power grid was plagued by problems. For one thing, no new generating plants had been built in recent years, but demand for electricity had soared. As a consequence, California had grown heavily dependent on out-of-state hydropower from the Pacific Northwest, and state legislators hoped that competition fostered by deregulation would strengthen in-state electricity generation and also bring down the cost of electricity for households and businesses.

California's deregulation laws were complex. Indeed, the buying and selling of electricity is endemically complex: it still can't be stored, the control of supply and demand needs to be exquisite at all times, and each set of transmission lines has limited capacity, which means that in a high-demand, high-volume market, the supply routes for all customers need to be charted out ahead of time to avoid congestion in the lines. To handle the buying and selling of electricity in the deregulated marketplace, California created an agency that set hourly prices for electricity, which could be bought at auction the day before or on the day of delivery. Another agency managed the transmission lines and conducted real-time auctions, which were meant to take care of last-minute, unexpected changes in supply and demand and guarantee adequate power reserves.

Initially, deregulation did reduce the cost of power. But the winter of 1999–2000 was a dry one in the Pacific Northwest, and the low snowpack meant that come spring, there would be less available hydropower from Oregon for the California utilities. In May 2000, unseasonably warm weather caused the demand for air conditioning to soar, and already scarce power sources became even scarcer. Within this tight market, some of those in a position to take advantage of unfavorable conditions did. Enron and other energy trading companies exploited loopholes in the California laws to create an appearance of even greater scarcity in the face of high demand. The energy traders weren't interested in simply acquiring energy; they made much of their money by "flipping" it — buying and then selling in order to make a profit on the trade. Their primary goal was not to provide better service to customers at a reasonable profit for themselves, but to make money.

The traders created artificial shortages by reserving energy for which they had no need and by inducing generators to shut down for maintenance even when maintenance wasn't necessary. They also scheduled large loads of electricity for limited transmission lines so that actual delivery of the electricity would be impossible. All these practices tied up supply and forced California utilities to buy energy at emergency prices during the real-time auctions. Wholesale prices of electricity increased more than 500 percent from the year before.

Retail prices during this time, however, were fixed, so utilities, including the state's largest — Southern California Edison and Pacific Gas and Electric — had to buy energy from the trading companies at a cost much higher than they could charge their private customers. The utilities, without the resources to purchase the electricity, pleaded for energy conservation and at times resorted to "rolling blackouts" in order to alleviate high demand they could not meet: one neighborhood after another went dark for an hour or two during the day.

Jeffrey Skilling blamed the shortage on poorly written legislation. "You probably couldn't have designed a worse system," he insisted. And while speaking at a conference in Las Vegas, he took the opportunity to joke about the dire situation in California: "You know what the difference is between the state of California and the *Titanic*?" he asked. "At least the lights were on when the *Titanic* went down." No less callous were the Enron energy traders, one of whom was recorded as saying, "They should just bring back fuckin' horses and carriages, fucking lamps, fuckin' kerosene lamps."

The crisis, which was alleviated only by state and federal intervention, ultimately cost California billions of dollars, and it induced other states to pull back from deregulating their power grids. Enron eventually failed spectacularly, Jeffrey Skilling ended up in jail, and new federal legislation aimed to reign in energy traders, but the California debacle served as proof that power greater than a million disciplined, unquestioning men was no match for greed.

Now, a little more than a hundred years after the first long-distance transmission lines at Niagara Falls sent power to Buffalo, more than 300,000 miles of lines capable of carrying more than a million megawatts of power form a caul over the country. And for all the blackouts and anxieties of the past half century, most of us still take our power on faith, for by now we might feel that we cannot see — or even think! — without it, and its humming is not only the music of our spheres but also a kind of cathedral tune. To cartographer Steven Watt, who studied aerial photographs as he worked on new maps to be used with GPS technology, the transmission lines that cut across the country took on the appearance of leaded strips supporting panes of glass:

> I used contemporary satellite and aerial imagery to help me correlate the position of roads with other features in the landscape. For maps in the United States, I worked one county at a time, within which I redrew the position of every road and intersection. . . . I looked for a fixed reliable set of points independent of the roads, which I could use to subdivide each county, and I decided to use the electric power lines, which, for the most part, are straight and clearly visible in overhead imagery. Because trees must be cleared beneath them, they often appear lighter in color than their surroundings. They cross roads, rivers, and towns, dividing the land into sharp-edged polygons that become smaller as population density increases. . . .
>
> As I finished my work within each polygon, I drew a line along the edge of the power lines until it closed the polygon, which I would then fill in with color. Then I'd reemphasize the power lines by drawing over them in black. As I continued working, this process had the unexpected and beautiful effect of creating a pattern reminiscent of a medieval stained-glass window.

Yet this enormous accomplishment, which the National Academy of Engineering has designated "the most significant engineering achievement of the 20th century," is also in dire need of reimagining. Not only is the grid no longer adequate for the increased demands of our times, but it suffers from age and neglect. The generating plants and transformers are old

— many are working beyond the natural end of their thirty- and forty-year life spans. Power companies routinely fail to attend to necessary maintenance, including tree trimming, along the lines. It's no coincidence that there have been five massive blackouts in forty years, and three of them have occurred in the past decade. Most significantly, in the United States coal constitutes the greatest single electricity fuel source, powering more than 55 percent of our electric plants (oil now accounts for less than 3 percent of the fuel used to produce the nation's electricity), and further exploitation of fossil fuels isn't tenable in a time of climate change.

The grid of the future, still more imagined than real, may take the shape the U.S. Department of Energy envisions: a system that relies on renewable power sources in the heart of the country. Large solar farms in the desert and wind farms on the Great Plains could produce power that would be delivered as far as the Atlantic and Pacific coasts, where the demand for energy is highest. In such a scenario, transmission lines will need to carry much heavier loads than existing ones can bear, and they will have to be much more efficient than the lines of today, which lose up to 7 percent of power during transmission. As far back as the 1990s, Richard E. Smalley, Nobel laureate and former director of the Carbon Nanotechnology Laboratory at Rice University, suggested that new transmission lines could be built of carbon nanotubes. These tubes are smaller than a blood cell, resilient, and capable of conducting electricity quite efficiently — more efficiently than copper or aluminum — which means that more energy can be transmitted over fewer lines with far less loss along the way. Still, the practical application of nanotube technology is not yet feasible, and the cost of building such a grid would be enormous.

The construction of such a system faces other challenges as well. For one thing, a grid that depends on solar and wind power needs to be able to compensate for the natural fluctuations of these energy sources. Proponents of a stronger national grid imagine that it will be a smart grid, with sophisticated monitoring systems that could match supply to demand across large areas. At times when the wind died down on the plains, the system could instantaneously redirect demand to, perhaps, solar generators elsewhere.

The smart grid would eventually be able to monitor home use and would work in concert with energy storage. Although electricity, except in very small quantities, still can't be stored effectively, Smalley imagined that all homes and businesses would have systems that could store a short-term supply of it — twelve to eighteen hours' worth. Such storage capability would be coupled with a system of real-time pricing, meaning that electricity would cost the most at the hours of greatest demand, so customers would have an incentive to avoid purchasing power during peak times. They could buy it as they slept to even out the demand on the system.

The national grid would also be accompanied by a strengthened localized system of energy production. In a smart grid, the meters would spin both ways, which would enable surplus from small local sources of power, even backyard windmills and solar rooftop panels on private homes, to be instantaneously sold back into the system.

Other energy specialists, including environmental writer Bill McKibben, envision a different grid of the future. They put greater store in applying smart grid technology almost solely to local power sources. There would then be little need to develop expensive long-distance power lines. Each area of the country would exploit the renewable sources — wind, tides, sun, falling water — that it had in abundance. McKibben envisions a decentralized, scaled-back, localized grid that would be seamless, intricate, finessed: "Imagine all the south-facing roofs in your suburb sporting solar panels. Imagine a building code that requires all new construction to come with solar roof tiles and solar shutters. Imagine windmills scattered around town in gustier spots and heat pumps for extracting energy from the earth. Imagine all these pieces linked in a local grid, supplemented with small-scale fuel-burning power plants that produce not just electricity but heat that can be pumped back out to local buildings."

It's one thing to recognize that the grid is no longer adequate, another to build a viable alternative. The development of nanotube technology, energy storage, and smart grid technology will require large-scale research-and-development projects and investment in science education, both of which will take considerable private and public funding. Smalley, before his death from leukemia in 2005, commented:

> Energy is at the core of virtually every problem facing humanity. We cannot afford to get this wrong. We should be skeptical of optimism that the existing energy industry will be able to work this out on its own.... America, the land of technological optimists, the land of Thomas Edison, should take the lead. We should launch a bold New Energy Research Program. Just a nickel from every gallon of gasoline, diesel, fuel oil, and jet fuel would generate $10 billion a year.... Sustained year after year, this New Energy Research Program will inspire a new Sputnik Generation of American scientists and engineers.... At best we will solve the energy problem within this next generation; solve it for ourselves and, by example, solve it for the rest of humanity on this planet.

It's not only the grid that needs reimagining. Homes and businesses need to become more energy efficient, too, as does lighting itself, especially now that many of us use more — and brighter — lights than ever before. Illumination still accounts for 6 to 7 percent of the energy consumed in the United States. We can create almost any effect we want: Ambient light can be diffused throughout the room. Bulbs can be recessed, shielded, layered, activated by sensors, or gradually dimmed. Lighting within a room can change hour by hour, with mood, with purpose. But in American homes, all these many effects are still largely achieved with incandescent light.

The most efficient practical cold light remains fluorescent, and in some respects fluorescent's quality has vastly improved since being showcased at the 1939 New York World's Fair. The delay time is shorter, the buzzes and flickers have diminished, and compact fluorescent lights (CFLs) can fit into traditional incandescent sockets. But it's also true that the general quality of CFLs has been erratic, especially in recent years, as companies have attempted to lower the purchase price of them. And they still aren't as versatile as filament bulbs. Some CFLs can't be used with dimmer switches, and the life of others is diminished when they are used in confined places such as recessed ceiling fixtures, which tend to get quite warm.

But fluorescent light remains coolly efficient. A new 13-watt compact fluorescent produces as many lumens as a 60-watt incandescent bulb while using one-quarter of the electricity. Its use eliminates more than a thousand pounds of global warming pollution. Because of such efficiency, CFLs have gained favor in many countries over the past few decades. British lighting historian Brian Bowers notes, "From about 1990 [compact fluorescents] were readily available in high street shops, and by 1995 half the households in Britain were using at least one." By the mid-1990s, half of the households in Germany used CFLs, as did more than 80 percent of households in Japan. In Asian countries in general, compact fluorescents are often more common than incandescent bulbs. One recent traveler to Korea noted, "It took me almost two months of living in Korea before I saw my first incandescent ('old-fashioned') light bulb. All of the others were energy efficient CFLs . . . [which] are so common here, in fact, that only in one store have I ever actually seen old-fashioned bulbs for sale, and that was in a dollar-store of sorts."

Yet CFLs are still not an easy sell in the United States: "You wake up and you're kind of groggy, and then you see these curly light bulbs, and it's buzzing, and you're like — ugh." At best, compact fluorescents prompt a soldierly acquiescence, wrapped in the anxieties of the age: "No, the light quality isn't ideal, and in some you can hear a slight buzzing . . . but I will have a hard time telling my children that I didn't do much to alleviate climate change because of aesthetics."

While developers of compact fluorescents continue to search for a more accommodating white light (read: closer to that of incandescence), the use of compact fluorescents has slowly accelerated in this country. In 2008 they constituted about 19 percent of all bulbs sold in the United States. Eventually, consumers may have little choice but to purchase them once new efficiency standards for illuminants imposed by Congress begin to take effect in 2012. These new standards will make the sale of most incandescent bulbs illegal. In response, researchers are currently developing more efficient incandescent lights, such as the Philips Halogená, but they are ten times more expensive than standard filament bulbs.

The efficiency of CFLs means that their use lowers mercury emissions at coal-fired generating plants, but CFLs themselves — like all fluorescents — contain mercury, a highly toxic metallic element that accumulates in the environment and can affect the nervous systems of living creatures. And at the moment, the disposal of compact fluorescents isn't regulated. Almost all compact fluorescents — and the mercury in them — end up in the trash, creating a considerable environmental problem of their own. In 2009 the state of Maine adopted the first extensive regulations concerning compact fluorescents, and once the law goes into effect, it will limit the amount of mercury manufacturers are allowed to use in the bulbs. The law also requires that a mandatory recycling program, paid for by bulb manufacturers, be established by 2011.

In the meantime, the Maine Department of Environmental Protection is so concerned about mercury from bulbs leaking into the environment that it not only urges householders to carefully recycle compact fluorescents but has also posted a fourteen-point instruction sheet on how to clean up one broken bulb. It begins: "Do not use a vacuum cleaner to clean up the breakage. This will spread the mercury vapor and dust throughout the area and could potentially contaminate the vacuum. Keep people and pets away from the breakage area until the cleanup is complete. Ventilate the area by opening windows, and leave the area for 15 minutes before returning to begin the cleanup. Mercury vapor levels will be lower by then."

The environmental problems posed by the mercury in CFLs is disadvantageous enough to brand them as transitional lighting, eventually to be replaced, perhaps, by light-emitting diodes (LEDs), which are composed of miniature plastic bulbs illuminated by the movement of electrons in semiconductor material. There is no filament to burn out and no mercury to recycle. They are the coldest of lights. LEDs are already used widely for digital time displays, scoreboards, traffic signals, and Christmas and other decorative lights. In the past few years, as the technology has advanced, they've begun to be used for street lighting and, more rarely, for interior lighting,

but the "white" light still has a bluish cast, and unlike traditional bulbs, LEDs shed light in one direction only. Although
it's possible for LEDs to last decades, they are still quite expensive to purchase — generally more than ten times the cost
of an incandescent bulb.

Major lighting companies such as General Electric and
Philips are already looking beyond LEDs to organic light-
emitting diodes (OLEDs), which work by passing electricity
through thin layers of organic semiconductor material that
is sandwiched between charged substrates. OLED lighting is
still in its research-and-development stage, but its champions
have faith that it will last ten times longer and "burn" ten times
more efficiently than incandescent lights. OLEDs, a true de-
parture from the past, are flat and emit light over their entire
surface, creating large areas of homogeneous illumination. Although in their current state they're rigid and look like a mirror when turned off, eventually the diodes will be embedded
in bendable plastic substrates, which will be transparent when
the light is off. The light itself will be flexible and will be able
to change its form. It could cover an entire wall or ceiling, or
be wrapped around a column.

Can a screen ever be a lamp? Will we take to light
everywhere and centered nowhere? Light freed from the limits of the socket, the tube, and the filament? Or will we feel
lost in the wash of abundance? Gaston Bachelard, who glori-
fied the intimacy between a solitary soul and a slight, disciplined flame, valued the conversation between a thinker, a
lamp, and a book because he saw the lamp as a "polestar" to
the page: one reads, then looks at the flame and dreams. The
dreaming and reading and thinking are intertwined, everything alive at once and encompassed within the reach of the
flame. "The candle does not illuminate an empty room; it il-

luminates a book," he wrote, and both light and words possess their own distinct time: "The candle will burn out before the difficult book is understood."

If today one were to come upon a single light in a dark room, it would likely be the blue and white flickers emanating from a computer screen: *our* window, where the pages change and change again with the tap of a keystroke — the new sound of solitude — and the mind flickers along the jetsam of information — news, weather, work, a remark from a friend, advice, purchases. Tap, tap, tapping and gazing forward. There is no polestar to the page, for there is no distance between the light and the letters, both of which emanate from the screen.

Soon now, the faint tinkling of a broken filament will become another sound of another century. But for the moment, stubborn devotees of Edison's light remain: some are already hoarding incandescent bulbs; others are purchasing replicas of early electric lights. One lighting catalog, which offers a wide range of compact fluorescents and encourages energy efficiency, also offers for sale reproduction nineteenth-century bulbs. Their ornate carbon filaments — shaped like cages, or dipping and turning — are reminiscent of those the visitors to the Menlo Park lab might have encountered. They are as dim as all lights of the past: the 1890 Bulb and the Caged Bulb, 40 watts; the Victorian Bulb, 30 watts. And you will pay dearly for such small light. The advertising copy notes that carbon filament bulbs offer "the unique combination of ⅓ the light at 10 times the price of standard bulbs, but they make any fixture ethereally beautiful."

Such stubborn fondness for the age of incandescence is more than simply nostalgia. It's testimony to how much incandescent light has meant, and how perfectly suited it still seems to be, to modern life: the steady, brilliant light of a speeding century; light born of invention but also warm (or so it has come to seem), versatile, dependable, and economical (and in the end, democratic); light that brought with it an entirely new world full of gleaming things; light at a far remove from whaling ships toiling in frigid waters and the stink and fuss of kerosene. It's also true that unlike kerosene — which began as the oil people had dreamed of for centuries and, within a few decades, ended as a symbol of exclusion from the modern — "old-fashioned" bulbs still shed a more satisfactory light than anything yet developed to replace them. And perhaps they always will.

At the Mercy of Light

EVEN IF WE CONSTRUCT a more resilient, sustainable grid that can meet the ever-increasing demand for more electricity, and even if we fully trade incandescence for the equivalent illumination in LEDs, we will still need to reimagine the accumulated brilliance we now think of as ordinary, for it turns out that the sheer abundance of artificial light, whatever its source, has consequences for our physical and spiritual well-being. And more than that: mammals, insects, birds, plants, and fish all find themselves at its mercy.

The understanding of the way artificial light adversely affects living things is still an unfolding mystery, but we do know that ubiquitous light wreaks havoc with our circadian rhythm — our daily cycle of variations in body temperature, hormone levels, heart rate, and sleep-wake times that is controlled by our biological clock. In humans, as in all mammals, the clock consists of a small cluster of nerve cells in the hypothalamus, which is cued by the varying levels of light that reach it from the retina. Having evolved in the absence of artificial light, it's broadly attuned to sunrise and sunset.

Researchers once believed that people living in modern industrial societies might have evolved away from the human biological clock as it functioned in earlier times, for the workings of our internal clock aren't always obvious, divorced as we are from the environmental constrictions of days, months, and years: we control our heat and light, and we no longer breed on a seasonal cycle. However, the experience of French geologist Michel Siffre, who in the summer of 1963 descended into the Scarasson Cave — a glaciated cavern under the French-Italian Maritime Alps — where he spent more than two months without sun, helped to confirm that our internal clock continues to keep time independent of our modern way of life.

Siffre set up camp alongside an underground glacier, surrounded by the corrosions and dissolutions of the cave, its dripping darkness and cold. Though he had one incandescent bulb to see by, he had no way of knowing the hour. "I wanted to investigate time," he explained, "that most inapprehensible and irreversible thing. I wanted to investigate that notion of time which has haunted humanity since its beginning." He reckoned his days by awakenings, recording each in a diary, and also telephoning scientists on the earth's surface, who registered the actual hour of his call, though in conversation they never told him what day it was or the time of day.

During his months alone in the dark, he had to cope with isolation and loneliness, and with the threats from the unstable walls and ceilings all around him. "This morning I was completely stunned," he wrote, after hearing a series of loud cave-ins of rock and ice. "My pulse was rapid, my mind full of dark thoughts. In such moments one realizes one's insignificance. . . . Birth, life, death, and then — nothing. No, no! Birth, *creativity*, and death — that sums up a man; the rest belongs to the animal kingdom. When I had partially recovered from my fright I looked at myself in the mirror: a pale and puffy face, with haggard eyes brimming with tears stared out of the glass."

The anxieties, confusion, and physical stress of those months would take their toll. "I emerged," he would say later, "as a half-crazed, disjointed marionette." Even so, he meticulously recorded his observation of time, and his diary is the record of a man who has lost all comprehension of duration:

> Forty-second awakening: . . . I really seem to have no least idea of the passage of time. This morning, as an example, after telephoning to the surface and talking for a while, I wondered afterward how long the telephone conversation had lasted, and could not even hazard a guess. . . . Fifty-second awakening: . . . I am losing all notion of time. . . . When, for instance, I telephone the surface and indicate what time I think it is, thinking that only an hour has elapsed between my waking up and eating breakfast, it may well be that four or five hours have elapsed. And here is something hard to explain: the main thing, I believe, is the idea of time that I have at the very moment of telephoning. If I called an hour earlier, I would still have stated the same figure. . . . I am having great difficulty to recall what I have done today. It costs me a real intellectual effort to recall such things.

During Siffre's months underground, the scientists on the surface keeping track of his daily cycles of waking and sleeping saw that they remained quite near a 24½-hour cycle: his internal clock had not shifted, only his conscious understanding of time. But Siffre parsimoniously meted out his rations to himself, for in misunderstanding the length of his day, he believed to the last that he had weeks more to endure. At the fifty-seventh awakening — the final day of the experiment — Siffre believed it to be August 20 when, in truth, it was September 14: the time graph he'd kept lagged twenty-five days behind the actual date. "I underestimated by almost half the length of my working or waking hours; a 'day' that I estimated at seven hours actually lasted on the average fourteen hours and forty minutes," he commented after his emergence from the cave.

Siffre's experience proved that our circadian rhythm may be able to withstand the periodic absence of light, but additional research since then suggests that even small amounts of artificial light can significantly disrupt that rhythm. The effects of artificial light on sleep are particularly profound, for it is the absence of light that induces our biological clock to signal the pineal gland to increase production of the sleep-inducing hormone melatonin. Although bright lights are difficult to separate from other things that may contribute to troubled sleep — noise, coffee, busy evenings — Dr. Charles Czeisler, who conducted a study of human response to light at Brigham and Women's Hospital in Boston, found that not only intense artificial light but also long periods of lower-level artificial light can disrupt the human biological clock. As a result, the clock can be shifted by up to four or five hours, "meaning that most people in the United States are actually on Hawaii time. Instead of people experiencing a peak drive for sleep between midnight and 1 A.M., for most people this is now at 4 A.M. or 5 A.M. . . . [They] are forced to wake up earlier than they would like to and remain tired during the day." Dr. Czeisler notes, "Every time we turn on a light we are inadvertently taking a drug that affects how we will sleep and how we will be awake the next day."

Additionally, in modern industrial societies, humans tend to give themselves little time to wind down in darkness and quiet before attempting to go to sleep. And they no longer vary their sleep according to seasonal changes in the length of days and nights, although even now the human biological clock still shifts according to the season and the amount of sunlight in a day. For instance, in the north temperate latitudes, the biological night is long during the winter and short during the summer, but people often bathe themselves in sixteen hours of light during all seasons of the year, as if every night fell during high summer.

Even the eight hours of uninterrupted sleep now considered desirable may be something imposed by industrial society, which requires every day of the year and all hours of the day to be divided in a certain way: now work, now relaxation, now sleep. Historian A. Roger Ekirch discovered that medieval villagers slept in a different way from modern people. Each night, they experienced divided sleep. They would go to bed soon after sundown, sleep for four or five hours — this was called "first sleep" — and then wake up an hour or two after midnight. Some people inevitably took advantage of the early-morning hours to get out of bed and work: students bent over their books; women did housework they couldn't get to during the day. Some even visited neighbors or slipped out of the house to steal firewood or rob an orchard. It was a good time for sex. But frequently people would lie quietly in bed, resting or talking, before they fell back into a lighter, dream-filled sleep — called "second sleep" — that lasted until sunrise. The quiet, free time in the small hours would have been dearly valued in a society where the days were filled with labor and obligation.

Divided sleep, Ekirch notes, began to slip away as artificial light increased. By the seventeenth century, the wealthy, who already prized their nightlife, no longer experienced it. Later, as the middle class acquired increased light, divided sleep slipped from them as well. Then laborers lost it, though vestiges of it remained even into the late nineteenth century. Robert Louis Stevenson, who sometimes slept in the open during his journey through the Cévennes in southern France, observed that a wakeful period in the middle of the night was a natural occurrence not only in people still living close to nature but in all of nature:

> There is one stirring hour unknown to those who dwell in houses, when a wakeful influence goes abroad over the sleeping hemisphere, and all the outdoor world are on their feet. It is then that the cock first crows, not this time to announce the dawn, but like a cheerful watchman speeding the course of night. Cattle awake on the meadows; sheep break their fast on dewy hillsides, and change to a new lair among the ferns; and houseless men, who have lain down with the fowls, open their dim eyes and behold the beauty of the night. . . . At what inaudible summons, at what gentle touch of Nature are all these sleepers thus recalled in the same hour to life? . . . Even shepherds and old country-folk, who are deepest read in these arcana, have not a guess as to the means or purpose of this nightly resurrection. Towards two in the morning they declare the thing takes place.

Given a chance, many humans will fall back into that medieval pattern of sleep, which may have been the way even the first humans slept. When Dr. Thomas Wehr and researchers at the National Institute of Mental Health attempted to replicate prehistoric sleep conditions by imposing on a group of men a daylight time of ten hours — what people in the middle latitudes during the dead of winter experience — he found that they

> slept only about an hour more than normal, but the slumber was spread over about a 12-hour period. They slept for about four to five hours early on, and another four to five hours or so toward morning, the two sleep bouts separated by several hours of quiet, distinctly nonanxious wakefulness in the middle of the

night. The early evening sleep was primarily deep, slow-wave sleep and the morning episode consisted largely of REM, or rapid eye movement, sleep characterized by vivid dreams. The wakeful period, brain wave measurement indicated, resembled a state of meditation.

"We think Thomas Edison had a bigger effect on the human body clock than anyone realized," remarked Dr. Czeisler. Edison, who favored catnaps on his laboratory tables, would have loved to think so, for he once commented, "Everything which decreases the sum total of man's sleep, increases the sum total of man's capabilities. There is really no reason why men should go to bed at all." Few would now agree with Edison, for although we may not yet know why we need to sleep, most people now understand it to be essential. As researchers look more deeply into sleep, they increasingly discover the true toll its lack takes on the physiological and psychological well-being of humans. Sleep-deprived people are more prone to elevated blood pressure and blood glucose levels. Lack of sleep depresses the immune system, affects memory and brain function, and shifts levels of the hormone leptin, which controls appetite, so it may also contribute to obesity.

We humans can alleviate the way artificial light creates havoc with our biological clocks: sleep institutes, sleep programs, sleep doctors all prescribe a regimen that re-creates ancient life. In addition to advising insomniacs to get daily exercise, avoid stimulants, and slow down in the evening, experts suggest that they avoid bright light at night, go to bed in a dark room, and sleep until daylight. But other creatures adversely affected by our light can do little more than suffer its effects or adapt to it. Nocturnal animals hunting in the dark, as well as those abroad in daylight that sleep at night — standing up, or with one eye open, or in hiding — are at its mercy, and human light not only affects their circadian rhythms; it can also compromise their chances for survival and even alter their evolutionary trajectories.

As with humans, the effects of ubiquitous light on wildlife aren't always easy to isolate, since accompanying them are countless other environmental changes and losses to habitat. Buildings and roads disrupt forage routes; noise and activity compromise many animals' ability to hunt; human endeavors create new ecosystems in which artificial light plays only a part. According to William A. Montevecchi, "Offshore hydrocarbon platforms develop rapidly into artificial reefs that create marine communities. These reefs attract, concentrate and proliferate flora, crustaceans, fishes, and squids. . . . Lighting attracts invertebrates, fishes, and birds, and organisms at higher trophic levels are in turn attracted to lower ones as well as to the lighting."

But light alone also changes everything. For many nocturnal animals, night is a negotiation between hiding and seeing. Mammals prefer to stay in the shadows and tend to avoid the full moon, which exposes them and makes them vulnerable to predators. Artificial light not only makes it more difficult for animals to hide; it also makes it more difficult for mammals that depend on keen night vision for both food and safety to see:

Many nocturnal species are using only the rod system, and bright lighting saturates their retinas. Although many . . . have a rudimentary cone system and can switch over to it within a couple of seconds, during those seconds they are blinded. Once they switch to the cone system, areas illuminated to lower levels become black, and the animal may become disoriented, unable to see the dark area . . . [ahead] and unwilling to flee into the unseeable shadows whence it came. . . . Finally, if the animal is in the lighted area long enough to saturate its rod system, it will be at a distinct disadvantage for 10–40 minutes after returning to darkness.

Light also changes the way nocturnal animals negotiate their world. A road lined with streetlights creates a kind of visual barrier: an animal cannot see beyond the lights and must take extra time, caution, and effort to make its way. One scientist, in studying the habits of pumas in southern California, observed that when a puma was "exploring new habitat for the first time, [it] stopped during the night at a lighted highway crossing its direction of travel. . . . In several instances, the animal would bed down until dawn, selecting a location where it could see the terrain beyond the highway after sunrise. The next evening, the puma would attempt to cross the road if wildland lay beyond or would turn back if industrial land lay beyond."

No less essential for creatures than the dark is the natural light of the night. Since light travels in straight lines, birds and mammals use celestial light for both navigation and orientation. When human light intrudes, it can misinform and confuse them. Consider the consequences of artificial light on birds. For centuries, nocturnal flyers have been drawn to lighthouses. Back when the Eddystone lighthouse keepers were eating their candles, the small light probably wasn't much bother. But a 1912 illustration shows the Eddystone lighthouse clouded by flocks of birds, milling and confused, streaming skyward, circling the white stone tower. In modern times, the dangers are multiplied: birds are drawn to the myriad illuminated windows of tall buildings and skyscrapers, and to the lights on broadcast and communication towers, which they either crash into or circle until they are exhausted. Birds also congregate around the flares on offshore oil and gas platforms, especially "on misty and foggy nights, and as they fly near and through the flames they are burned to death." And not just elevated lights cause problems: water birds and marsh birds can mistake light-reflecting surfaces for water, and once they land on dry ground, they can't easily take off again. While they struggle to fly away, they remain exposed and vulnerable. And nocturnal seabirds that hunt for bioluminescent prey are mistakenly attracted to lights and confused by them; as a result, their search for food is frustrated. In all cases, if birds that are trapped by light manage to escape death, they've expended precious energy they can't afford to waste. For those in migration, their confusion often delays their arrival at breeding or wintering grounds.

It is not only light itself but the duration of the light that affects birds. They, too, are exhausted by sixteen-hour days. Artificial light triggers their dawn response and leads them to sing after sunset, sometimes to sing all night. The artificially extended day affects their migration and breeding patterns as well.

In the animal world, even what goes on under one street-light after dark has complex and far-reaching consequences, for a single light is capable of changing the equilibrium of an ecosystem. Moths and insects gather around a streetlight; bats and toads come in to pick at easy prey. Notes one scientist, "The habit of feeding at artificial lights is now so common and widespread among bats that it must be considered part of the normal life habitat of many species." This not only increases the stress on insect populations; it also changes the relations between different bat species, since not all species use lights for feeding, though they may feed on similar insects. The presence of streetlights gives species that use lights a competitive advantage over other species. The non-light-using species may decline because they have lost their competitive edge. By altering habitat and spurring adaptations that might eventually become encoded in the future lives of insects, mammals, birds, and reptiles, "humans are changing the evolutionary trajectories of those affected species, causing them to adapt to new sets of conditions," notes biologist Bryant Buchanan. "Simply conserving species richness or population sizes does not conserve the evolutionary and behavioral diversity contained in those taxa."

Sometimes artificial light becomes an evolutionary trap as the age-old biological imperatives of a species, which helped it survive for eons, turn into liabilities. The most well-known example of such a trap is the predicament of the loggerhead turtle, which can live for more than 130 years. It inhabited coastal waters long before humans existed on earth, trolling the shallows, feeding on sand dollars, whelks, and conchs. The female, year after year, crawls out of the surf and onto sand beaches to nest. She has always preferred the cover of darkness for safety, and now the bright lights of shoreside developments often drive her away from prime nesting sites. When she does settle on a nesting place, she digs a pit along a sandy shore with her flippers, then deposits a clutch of eggs, falling, "as they have fallen for a hundred million years," writes David Ehrenfeld, "with the same slow cadence, always shielded from the rain or stars by the same massive bulk with the beaked head and the same large, myopic eyes rimmed with crusts of sand washed out by tears."

She then covers the nest and returns to the sea. The eggs take months to develop, during which time, if the female has not been able to nest in the best of places, they are all the more vulnerable to extreme high tides, storms, and predators. If they survive their incubation period, the hatchlings then extricate themselves and dig their way — en masse — to the surface. If the surface sand is hot, they know it to be daylight, and they burrow back down and wait until the sand cools after sunset. Then they begin their trek to the sea. They are keyed to move toward the lightest horizon, and for thousands of years this meant they crawled away from dark dunes and vegetation and toward the ocean, whose surface, glinting and sparkling with reflecting starlight and moonlight, was brighter than the interior land. In a dark landscape, the baby turtles usually have no more than a two-minute trip to the beach.

But on developed beachsides, lit with condominiums, streetlights, and commercial districts, the turtles are confused by the brilliance of the built landscape at night. They crawl toward high land instead of the sea; crawl into roadways, where they are killed by cars; or crawl so far that they die of exhaustion. If they manage to reorient themselves and somehow reach the water, their mortality rate — already considerable, for they have to breach a surf rife with predators and then swim for at least a full day to reach their dwelling grounds — is much higher.

Amid the brilliance, it seems almost nothing remains unaltered by light. It affects the foraging and schooling patterns of fish and the timing of migrations. It alters the drift stream of insects on water and the vertical migration of zooplankton and fish. It diminishes the effectiveness of bioluminescent creatures. Fireflies were once bright enough to light up a village night. Now human light washes out their glow, which makes it harder for them to attract mates. And plant life is not immune. Measured light and darkness signal plants that the right pollinators are available and that competition is minimal. The coarse and prickly cocklebur (*Xanthium pensylvanicum*) — thriving in vacant lots and dumps, catching on clothes, riding on fur — flowers nevertheless and is keyed to its optimum bloom time by the length of the night. But the dark needs to be continuous: "A light break as short as one minute in the middle of a long night would prevent [it] from flowering."

Even when our lights are meant to be their most heartening and consoling, they have consequences for wildlife. In 2004, during the annual Towers of Light tribute to commemorate those killed on September 11, 2001, spectators in New York City wondered at "the thousands of little stars . . . suspended in the air." It was a calm, moonless night during the fall migration. The upward flow of warm air in the columns of light induced moths to circle in the lights for fifteen stories or more, and thousands of birds, also drawn to the columns, circled above the moths. Few people understood what they were seeing. "Some people thought they were specks of dust," reported the *New York Times*. Others, perhaps remembering the rain of debris on that clear day three years before, concluded otherwise: "Some people saw ashes. Some thought there were fireworks in the light columns. Some saw spirits."

What they could not see, of course, were the actual stars, most of which were obliterated by the brilliance of the city night.

MORE IS LESS

At the second match the wick caught flame. The light was both livid and shifting; but it cut me off from the universe, and doubled the darkness of the surrounding night.

—ROBERT LOUIS STEVENSON,
Travels with a Donkey in the Cévennes

URING THE LATE NINETEENTH CENTURY, Vincent van Gogh saw countless subtleties in the dark skies of southern France: "One night I went for a walk by the sea along the empty shore," he wrote to his brother, Theo, in 1888. "The deep blue sky was flecked with clouds of a deeper blue than the fundamental blue of intense cobalt, and others of a clearer blue, like the blue whiteness of the Milky Way. In the blue depth the stars were sparkling, greenish, yellow, white, pink, more brilliant, more sparklingly gemlike than at home — even in Paris: opals you might call them, emeralds, lapis lazuli, rubies, sapphires." As van Gogh — aided by gaslight — painted that sky, he also painted myriad relations between the celestial and the human. In *Starry Night*, the illuminated village appears intimate — and inconsequential — against the roil of stars and the quarter moon above it, while in *Starry Night over the Rhône*, the human light and starlight are in conversation with each other: a couple stand at the lower right of the painting, and all around them the world is alive with light. Just beyond them, the river is ribboned with the reflection of the streetlights of Arles in the distance. And beyond the river, the town itself spangles the horizon. But it isn't too bright to stop the stars overhead or the sense of night as enormous and other. The night sky, defined by the brilliance of the stars, occupies almost half of the canvas.

Even in the midst of Arles, in *The Café Terrace on the Place du Forum, Arles, at Night*, human life negotiates a middle distance between the cobbled street and the stars. The glow of gaslight washes the walls of the café and its canopy roof; here and there a private, ruddy luminescence shines from second- and third-floor windows, and a few shop windows glow. But beyond the terrace, the dark increases quickly, and stars glitter in the gaps between buildings. Present-day astrophysicist Charles Whitney suggests that van Gogh "has overpopulated the small patch of sky in view of the interference that might be expected from the café lights." And van Gogh himself once insisted, "*I should be desperate if my figures were correct.* . . . I do not want them to be academically correct. . . . My great longing is to learn to make those very incorrectnesses, those deviations, remodelings, changes in reality, so that they may become yes, lies if you like — but truer than the literal truth." You can imagine, in the truth of his time, that even in the midst of nightlife, contemplation of the stars was part of being at home in the world, a counterpart to earthly life.

For many people, light pollution is now so pervasive that it obliterates any chance they may have to observe the night sky. In particular, sky glow — the orangey brightness in the air around cities, towns, and industrial sites that fades to purple in the upper night sky — hinders our seeing. Although sunlight reflecting off the moon, earth, and cosmic dust, and starlight scattering through the atmosphere, make for some natural sky glow, the ubiquitous light shed from homes, businesses, and streetlamps causes most of it. In the twenty-first century, even many wide suburban backyard views of the heavens have shrunk to a sprinkling of dim stars, and most people in the developed world see the night sky as if it is always washed in moonlight, at least as bright as a first-quarter moon. To people in large modern cities, the night sky always appears brighter than on nights near the full moon in the countryside, and the Milky Way — that bridge across the sky of dust and stars and gas, "brilliant with its own brightness," Ovid once wrote — can't be seen with the naked eye by two-thirds of Americans and half of all Europeans.

The Milky Way had always been the stuff of legend, variously called the Deer Jump, the Silver River, the Straw Thief's Way, the Way of the Birds, the Way of the White Elephant, the Winter Way, and the Heavenly Nile. It guided pilgrims at night and so was known also as the River of Heaven, the Road to Santiago, and the Roman Road. Now its appearance has become so unfamiliar that when the lights went out in Los Angeles during a 1994 earthquake, "emergency organizations as well as observatories and radio stations in the L.A. area received hundreds of calls from people wondering whether the sudden brightening of the stars and the appearance of a 'silver cloud' (the Milky Way) had caused the quake."

If you can't see the Milky Way anymore, you can't see a fourth-magnitude star, magnitude being the measure of how bright a star appears from earth. The brightest objects have negative magnitudes: the magnitude of Sirius is −1.4 and that of Venus is −4.5. In moderately light-polluted skies — where the Milky Way no longer appears — about three hundred second- and third-magnitude stars are still visible, but all the lesser ones — almost seven thousand fourth-, fifth-, and sixth-magnitude stars — are lost. Also, all the stars in light-polluted skies are less apparent than they were to our ancestors because the lights we live by are often so bright they suppress the rod system of the human eye: "About one-tenth of the World population, more than 40 percent of the United States population and one sixth of the European Union population no longer view the heavens with the eye adapted to night vision, because of the sky brightness."

The disappearance of stars is most keenly felt by astronomers, the true descendants of Galileo, who turned the first telescope toward the night sky. "Surely it is a great thing to increase the numerous host of fixed stars previously visible to the unaided vision," wrote Galileo in 1610, "adding countless more which have never before been seen, exposing these plainly to the eye in numbers ten times exceeding the old and familiar stars." His first observations, which included the discovery of four moons orbiting Jupiter, reinforced his belief in a sun-centered universe:

> Here we have a fine and elegant argument for quieting the doubts of those who, while accepting with tranquil mind the revolutions of the planets about the sun in the Copernican system, are mightily disturbed to have the moon alone revolve around the earth. . . . But now we have not just one planet rotating about another. . . . Our own eyes show us four stars which wander around Jupiter as does the moon about the earth, while all together trace out a grand revolution about the sun.

Galileo also observed that the moon Aristotle had perceived as perfect "is not robed in a smooth and polished surface, but is in fact rough and uneven, covered everywhere, just like the earth's surface, with huge prominences, deep valleys, and chasms." As for the Milky Way, he said: "With the aid of the telescope this has been scrutinized so directly and with such ocular certainty that all the disputes which have vexed philosophers through so many ages have been resolved, and we are at last freed from wordy debates about it. The galaxy is, in fact, nothing but a congeries of innumerable stars grouped together in clusters."

In the centuries following Galileo, as telescopes became more powerful and refined, astronomers increasingly saw farther back in space, and farther back in time — to Andromeda, to quasars and black holes — and among the optimum places for observing the stars were the higher elevations of southern California. The nights are generally clear there, and the mountains are not so high that their summits are lost in clouds or snow squalls, yet they rise above the dense atmosphere and fog of the coastal plain. The air is usually calm on the peaks as well: the prevailing onshore winds of the Pacific flow smoothly over them. This stability makes for what astronomers call "good seeing," for it is the movement of air flowing over the earth that distorts the light and causes the stars to twinkle. (By contrast, stars viewed by astronauts in orbit appear steady, while human lights on earth glitter.)

So exceptional was the seeing atop the peaks of southern California that during the first half of the twentieth century, the area became home to some of the most important observatories in the world. The first was the Mount Wilson Observatory, built in 1904 in the San Gabriel Mountains of Los Angeles County. "Many astronomers thought that on a good night the atmosphere over Mount Wilson was so still, the images of the stars so well defined, that it was perhaps the best seeing in the world," wrote historian Ronald Florence. But by the late 1920s, when George Ellery Hale began searching for an appropriate site to situate the 200-inch telescope he was to build, the city of Los Angeles and its suburbs had spread right to the base of Mount Wilson, and urban light was already compromising dark-sky work there. Consequently, Hale decided to house his telescope farther away from the cities, in a fern meadow on Mount Palomar, 5,600 feet above sea level. Palomar was still accessible, yet at forty-five miles from San Diego and one hundred miles from the Los Angeles basin—the 1930 census put the population of San Diego County at about 210,000 and that of Los Angeles and Orange counties at less than 250,000—it seemed safe from the effects of light pollution.

Hale and his backers decided where to situate the telescope in 1930, but it took almost two decades for the mirror to be completed—several years alone for it to be successfully cast of Pyrex at the Corning glass factory in New York and another year for it to slowly cool in an annealing oven, after which it journeyed by train across the country, moving at 25 miles per hour during the daylight hours and stopping after dark. Sixteen days after leaving the Corning factory, it arrived in a Pasadena, California, optics lab, where it remained for more than a decade as technicians, working with slurries of abrasives and with polishing rouge, ground away ten thousand pounds of glass and shaped it into a paraboloid. Meanwhile, crews improved the road to the summit of Mount Palomar, ran water and electric lines up the mountain, and built a dome to house the telescope. The Japanese attack on Pearl Harbor in 1941 put a stop to all work there, while almost everyone involved with the project was taken up by the war. The mirror was finally trucked up the mountain in 1947. Although the population of southern California had grown markedly and New

Deal electrification initiatives had increased the amount of light in homes and on the streets, Palomar remained a remote mountain rising out of the desert. Cattle grazed in the high meadows, and no appreciable light affected the observatory.

The Hale Telescope saw first light in January 1949, and on that occasion the eminent astronomer Edwin Hubble claimed: "The 200-inch [telescope] opens to exploration a volume of space about eight times greater than that previously accessible for study. . . . The region of space that we can now observe is so substantial that it may be a fair sample of the universe as a whole." After months more of adjustments—opticians polished the last five- or six-millionths of an inch of the mirror with hand-held cork tools and then their own thumbs—the telescope was officially turned over for exploration and research. Astronomers identified stars and studied their birth, evolution, and death; studied the workings of the galaxies; and searched for the age of the universe itself. "Astronomy is an incremental science," Florence wrote. "Each night adds data, fragmentary glimpses and measurements of the reaches of the universe. . . . Amidst that steady accumulation of knowledge, the achievements of the [Hale] telescope stand out as a history of twentieth-century astronomy."

But by the 1960s, light pollution began to compromise the quality of dark-sky study at Palomar—as it has at many observatories throughout the world in the past fifty years. In some places where dark-sky study is severely limited or impossible, institutions such as the David Dunlap Observatory outside Toronto and the Yerkes Observatory outside Chicago have been transformed into historical sites and education centers. Even in working observatories, quite a few celestial objects simply aren't apparent anymore. "It's like I'm looking for the glitter of a little pen-ray flashlight in the glare of bright sunlight," commented one astronomer at the Kitt Peak National Observatory outside Tucson, Arizona. "That 20 percent increase in sky brightness means it takes us 40 percent longer to record the same faint, distant objects. You get less done per expensive hour of operating the telescope."

Sometimes there isn't enough dark time in a night to record an object at all anymore. In addition, mercury vapor lamps — the most popular type of street lighting — do more than blot out the stars. They also compromise astronomers' ability to take an object's spectra — that is, split the light from the telescope into its component colors. Astronomer Dave Kornreich explains:

> When you take a spectrum of fluorescing objects like galaxies, you see that the spectrum is not smooth, but made up of a number of lines. Each line is a unique indicator of the presence of a certain chemical. By studying the strengths of these lines, astronomers can deduce the chemical composition and temperature of the objects they observe. . . . Mercury vapor lamps have an enormous number of these spectral lines in all parts of the spectrum [which] interfere with astronomical observations.

By 1980, when the population of San Diego County had grown to just under 2 million, light pollution around Palomar had become so severe that in the following years, scientists at the observatory, in an effort to counter the further increase of pollution, began working with the surrounding town and county governments to try to reduce unnecessary lighting and glare in the area. Light pollution may appear to be as complex as modern lighting itself, for not only do countless individual lights contribute to it, but different kinds of artificial light — incandescent, low- or high-pressure sodium vapor, low- or high-pressure mercury vapor, tungsten-halogen, fluorescent, and LED — affect the surroundings in different ways. And no matter the kind of light, the effect of lighting in any one place is always variable because its intensity and apparent brilliance are affected by weather, by the dust and gas in the atmosphere, and by the cloudiness or clarity of the sky. The direction and path of the illumination makes a difference as well. "Light traversing a path at a shallow angle above the horizontal . . . will cause more sky glow, since it will encounter more particles and droplets from which to be scattered on its way through the atmosphere," wrote astronomer Bob Mizon. The type of surface the light eventually falls on matters as well: whether it's wet or dry, smooth or rough, dark or light determines the reflectivity of any light.

The Palomar scientists and town and county officials attempted to mitigate the light problem at the observatory by creating zoning ordinances. For decorative lights, such as those used to illuminate advertising, and lighting at outdoor sales areas, they established strict shielding requirements, which would direct the artificial light downward. They also established an 11:00 P.M. curfew for nonessential lighting, and in the fifteen-mile radius around the observatory, ordinances forbade decorative lighting altogether. Riverside County replaced its mercury vapor streetlights with more efficient sodium vapor lamps, which don't interfere with the spectra of astronomical objects.

Even with concerted efforts to contain light pollution in southern California, the seeing at Palomar is becoming more and more deeply compromised. The mercury vapor lights from surrounding cities have grown so bright that astronomers can no longer observe some parts of the spectrum of celestial objects. In 1999 one astronomer noted that "the city lights can be directly seen through gaps in the mountains, meaning that city light is making its way directly into the telescopes, without first even being reflected by the sky. Many observers have given up looking at objects in the southwestern sky, because the light pollution is so bad in that direction."

As much as light pollution obscures an understanding of deep space, it obscures an understanding of time, for seeing outward "is equivalent to looking backward in time," Richard Preston notes.

> The universe — as we see it — could be imagined as a series of concentric shells centered on the earth — shells of lookback time. The shells closest to the earth contain images of galaxies near us in time and space. Farther out are shells containing images of remote galaxies — galaxies as they existed before our time. Still farther out is the shell of the early universe. Some of the photons reaching a telescope's mirror are nearly as old as the universe itself. The quasars are brilliant pinpoints of light that seem to surround the earth on all sides, shining out of deep time. Beyond the quasars, the observable universe has a horizon, which could be imagined as the inner wall of a shell. This horizon is the limit of lookback time, which is also an image of the beginning.

The Hubble Space Telescope — the first space-based observatory — orbits beyond the distortions of the atmosphere and the effects of light pollution, and it sends back to Earth clearer and deeper views of the universe than previously possible. But space-based telescopes — extraordinarily expensive and precarious — can't fully replace what happens when human thought meets the dark sky, whether at rarefied Palomar, where scientists gather and play cowboy pool as they wait for their precious observation time; in a remote pasture where an amateur astronomer collects starlight with a homemade telescope built of wood and mirrors; or at the back door of a farm-house, where a child looks up and stares.

Perhaps what we've lost with the disappearance of the night sky is more profound than we can possibly know. "Then Humankind was born," Ovid wrote in *Metamorphoses*.

Either the creator god, source of a better world, seeded it from the divine, or the newborn earth just drawn from the highest heavens still contained fragments related to the skies, so that Prometheus, blending them with streams of rain, molded them into an image of the all-controlling gods. While other animals look downwards at the ground, he gave human beings an up-turned aspect, commanding them to look towards the skies, and, upright, raise their face to the stars. So the earth, that had been, a moment ago, uncarved and imageless, changed and assumed the unknown shapes of human beings.

THE ONCE AND FUTURE LIGHT

The spiritual instant that is our life ...

—HENRI FOCILLON,
The Life of Forms in Art

W E IMAGINE THE FAR PAST as a world carved out of the dark, guttering flame by guttering flame, a past full of other people, and the crackle and smell of other lights: rushes, moss, spruce twigs, tallow, whale oil, and pine. Light in a world lacking it, where abundance and brilliance are a dream flaring for a moment and then gone, where light itself means one thing, the materials of it another: the rushlights of the poor, the beeswax of the church. But they aren't so far away, those people back there. We who live in a world built of light still carry the longings of those without it, still dream of abundance and brilliance even though we can have all the light we want — almost any kind — and have it in an instant.

Given what we have, and given what we know about the power of light and the limits of our resources, about the trajectory of a changing climate, how will we choose to illuminate our future? Can we overcome the desire for ever more and brighter light? Can we think rationally about light and what it means to us? One of the first things we ask of light is that it grant us some assurance in the dark. Except during the threat of aerial bombardment or under the glare of interrogation, it has almost always made us feel safer. But whether or not it truly ensures our safety is an open question, one that has been argued since the seventeenth century, when a few European cities expressly forbade streetlights for fear that they encouraged footpads and drunks, even as other cities were installing them in hopes of bringing order to the night.

Although criminals have historically avoided light — at least since the Middle Ages, when thieves shied away from nights of the full moon — it doesn't certainly keep them away when they sense an opportunity. British astronomer Bob Mizon notes, "The *Home Office Crime Survey*, published in October 2000 and based on the experience of victims of crime, suggests that premises which have security lighting are as likely to be broken into as those without it." Mizon also relates the story of "a car storage area, unoccupied at night, but not much lit up. It was near a major highway, and burglars would pull off, cut a hole in a fence, grab parts and leave, fast. The police finally caught one, and asked 'Would better lighting help?' The burglar replied: 'Sure, I could get in and out a lot faster and not get caught.'"

How complex the relation between light and street crime may be is illustrated by a study undertaken by the Illinois Criminal Justice Information Authority. In the late 1990s, researchers evaluated the impact of increased lighting in Chicago alleyways where the Department of Streets and Sanitation had replaced 90-watt lights with 250-watt lamps. In the ensuing months, violent crime at night increased by 14 percent, crimes to property increased by 20 percent, and substance abuse violations increased by 51 percent, while daytime offenses in the alleyways decreased by 7 percent. The authors of the study came to no clear conclusions as to why there was such a surge in crime, but they suggested that perhaps both citizens and police were witness to more offenses in the brightly lit alleyways, and consequently reported more of them. Perhaps, too, the greater illumination made residents feel more secure, so more people ventured out after dark, and the increased activity may have led to an increase in crime.

The correlation between light and safety may never be fully explained, for what light can do and what we imagine light can do are not entirely separate things. Our insistence on bright night lighting is inextricably linked with our feelings of safety, but those feelings are relative to our accustomed surroundings. Michel Siffre, living with one small light deep in his glaciated cavern, felt assured by his meager illumination. "Yes, my tent became my universe," he wrote. "Its effect upon my mind was remarkable. When I left the lightbulb on and went outside, the tent glowed in the cold darkness with a redness that was singularly comforting. From the moraine I often looked back at it with a feeling of love. It represented security and shelter — no matter how specious that security and shelter, which was threatened constantly by the cave-ins of rock and ice."

Even over the course of our own lives, the amount of light that makes us feel safe is a moving frontier: the more light we're accustomed to, the more we feel we need for security. For many of us now, abundant artificial light, not darkness, feels natural after the sun goes down. We not only walk in bright light; we also leave it behind us. Our houses are lit when we go out at night, and in the deserted small hours, light dazzles at rural crossroads, gas stations, empty parking lots, and shut-up summer homes and hotels. We turn our lights on in the early dusk and leave exterior house lights burning while we sleep. The lamps that assured our ancestors in the gaslight era would not be enough for us today.

Given the history of light's steady increase, nothing less than a conscientious, international effort will ensure a future in which the brilliance we live by stays the same, or — as astronomers David Crawford and Tim Hunter hope — markedly decreases. Crawford and Hunter were among the first voices calling for a return to darker nights, and in 1988 they established the International Dark-Sky Association for the express purpose of abating light pollution and increasing public awareness of the consequences of excess light. The association suggests strategies to reduce lighting — simply shutting off unnecessary lights would go a long way toward achieving this goal. (In the United States alone, wasted light costs more than $1 billion a year, and a 100-watt bulb left burning through all the dark hours in a year creates about five hundred pounds of carbon dioxide.) The group also promulgates the control of light through shielding and by directing essential light so that it illuminates only what it's intended to. For any new lighting, the association advocates extensive planning that takes into account an understanding of how illumination affects the surrounding environment.

In twenty years, the influence of the International Dark-Sky Association has grown far beyond a circle of stargazers, having gained the support of architects, city planners, and lighting designers:

A growing number of businesses are rethinking their attitudes toward the environment. . . . This shift has had some influence on the business of government, as well, including how we plan our day and night urban landscape. . . . Sustainability's emphasis in urban design and artificial night lighting is to improve the quality of lighting, not its quantity. [A more] holistic view of lighting design produces less environmental impact than poorly designed, more traditional approaches: it requires less energy from power plants. . . . And, ultimately, there is a net economic savings. . . . Because most of the Earth's population lives in cities or urban centers, nighttime lighting needs to be one of the key components of any city policy for urban development and for increasing the quality of life for its citizens. . . . While community planners remain firm in their mandate for safety, utility, and ambiance, some are beginning to examine the myths of night lighting, the meaning of utility, and the requirements necessary to create an effective and efficient ambiance.

Such rethinking is apparent in recent changes in the New York City skyline, where subtle and complex patterns of light, rather than mere brilliance, have begun to emerge. In part, the new pattern is a return to the old. In 1925 a writer for the *New York Times* declared that there was

a new city of light and color rising above an old one. . . . The illuminated towers of Manhattan are fast multiplying and the application of floodlights to their summits has brought about a fascinating aspect of architectural art. If the practice continues the glory of the cloud-hung castles of Camelot will pale before the reality of the illuminated citadels, towers, pinnacled turrets and minarets that even now rise above the city streets. . . . Crowning [the Standard Oil Building] is a pyramid illuminated by four huge flares, a beacon that is visible for miles at sea. . . . The Metropolitan Tower, with its red and yellow light clusters and its illuminated clock . . . may be read alike by deckhands on East River craft and watchers on the Palisades.

The advent of fluorescent light not only increased the light emanating from skyscrapers; it also changed the appearance of them at night. Once banks of fluorescents on the ceilings of offices remained on all night — long after even the cleaning crews had departed — a skyline shaped by scores of entirely illuminated tall buildings, of which the lit crowns were only a part, emerged: a skyline endlessly photographed and imagined, one that seemed to embody twentieth-century brilliance and electric energy. But in recent years, with the advent of energy-efficient lighting controlled by motion detectors, dimmers, and timers, and with ceiling lights that can be divided into zones, lighting designers can rely on more subtle effects — akin to what seemed marvelous in 1925 — to maintain the individual and iconic appearance of any skyscraper. One lighting designer has observed: "The tall tower with the illuminated floors on all night long is probably a thing of the past. You're not relying on the glowing floors to [give] the building presence . . . [but] on the crown of light." That crown of light may be illuminated by LEDs rather than floods, and it may even be more modest than the lit crowns of the old city.

Dimming and shielding lights not only increases sky darkness but also helps birds, mammals, and insects trying to navigate the night. Although the strategies for alleviating changes in wildlife habitat are complex, since even shielded streetlights can change the habits of bats and insects, the simplest of things can make an enormous difference in mortality rates, especially of birds. Chicago is situated along a major flyway, and during the spring and fall migrations, more than 5 million birds — at least 250 different species — cross the city skies. In past years, at night, many migrants either crashed into the illuminated buildings or circled them until exhausted. Every morning, the managers of Chicago skyscrapers would pick up dead birds by the shovelful from the roofs. Then city planners instituted the voluntary Lights Out Chicago program. They asked building managers to dim or turn off decorative lighting late at night and to minimize the use of bright interior lights during migration season — from mid-March to mid-June, and again from late August to late October. They also encouraged high-rise residents to draw their shades or dim interior lights late in the evening. As a result, bird mortality dropped by an estimated 80 percent.

Granted, in the heart of any major city, the night sky will never be dark, but damping down city and suburban lights to where they were even a few decades ago will bring the dark sky closer to more people. Astronomer John Bortle, creator of the nine-level Bortle scale, which measures light pollution, noted in 2001, "Unfortunately most of today's stargazers have never observed under a truly dark sky, so they lack a frame of reference for gauging local conditions. . . . Thirty years ago one could find truly dark skies within an hour's drive of major population centers. Today you often need to travel 150 miles or more."

To help stargazers gain a frame of reference and help in the preservation of knowledge that only the deep night offers, the International Dark-Sky Association has been working to create a series of dark-sky preserves — places far away from development and its attendant human light — where people can travel to see the pristine night sky. In the United States, these are often located at national parks in the darkest quadrants of the country. At one of the few preserves in the East, at Cherry Springs State Park in north-central Pennsylvania — more than sixty miles from the nearest city and sitting atop a 2,300-foot mountain — more than ten thousand people a year pay $4 to stand in an observation field in the middle of the park and be stunned by what was a common sight for people a century ago: the shadows cast by the Milky Way, the sheer number of stars.

In moderately light-polluted skies, which is the ordinary view for most people in the developed world, the major constellations, such as Orion, the Big Dipper, and Cassiopeia —

many of which are defined by second- and third-magnitude stars — stand out among the visible stars. In the darkest skies, what we know as the common constellations recede back into the multitude of stars. "There's a good part and a bad part," one amateur astronomer said of looking at the ten thousand stars visible at Cherry Springs State Park on a clear night. "It's good because there are so many stars. It's bad because there are so many stars. It's hard to keep yourself oriented sometimes."

But given time, those ten thousand stars would seem natural enough. Whether we are oriented by the constellations or not, to be in the presence of a truly dark sky is an unforgettable feeling: the stars have a palpable presence; you can almost feel the pressure of them.

As great a challenge as it is, reducing our light is only part of the solution, for a third of the world still isn't tied to an electric grid, and elsewhere grids are insufficient for demand — generators may be old and hydropower plants may not be able to provide consistent power. Although some societies continue to thrive with traditional lighting, in a more widespread electrified world, many without electricity feel its absence intensely, just as farm families in rural America did during the 1930s. In our intricate, interdependent global economy, where people living in vastly different circumstances have almost no ecological distance between them, simply creating sustainable industrial economies will mean little if the standard of living in less developed countries doesn't rise. Secure people will be far more interested in building sustainable economies for themselves than will those who must struggle to survive.

The lack of adequate light alone threatens to leave many people behind. In the country of Guinea, on the west coast of Africa, electricity generation has actually declined in recent years because of a deteriorating political situation. At its best, the country's hydroelectric power resources serve about 60 percent of its citizens, and then mostly in the rainy season and for only part of the day. Those in the countryside often have no electricity at all, so some country schoolchildren look for alternatives to candlelight. "When my mother buys me a candle at home to study, it doesn't last long," one student said. Some of them walk to gas stations near their homes to study under the outdoor lights; others camp in the yards of wealthy homeowners, reading with the help of the exterior lights and the glow of windows. Those who live within an hour's walk of the airport in the capital city of Conakry study in the airport parking area, amid the departures and arrivals of international flights, the roar of engines, the bustle of people coming and going. Older students sit on concrete pilings, bent over their notes, the fluorescent lights above them. Younger students hunker on curbs and traffic islands. "I hardly ever take notice of the arrival of planes or cars. . . . I am here to study," one student remarked. Another said, "I used to study by candlelight at home but that hurt my eyes. So I prefer to come here."

Although light alone won't change everything, bringing it to places such as Guinea will ensure more than illumination. The darkness can be mitigated with solar flashlights and lanterns, as well as other innovations that allow adults to work and travel after dark and children to study. For the Huichol Indians of Mexico's Sierra Madre Occidental, who live burrowed deep in rugged, sparsely settled terrain, new lighting technology has made a distinct, practical difference in their lives. Architect Sheila Kennedy, in an attempt to alleviate the Huichol's isolation from modern illumination, devised the Portable Light system, which harvests and stores solar energy during the day, then offers up to eight hours of light at night. It consists of a rectangle of fabric with LED chips attached to one side. The fabric, coated with an aluminum film, reflects the light produced by the diodes. On the reverse side, two flexible solar panels are sewn onto the fabric, and these power a lithium battery stored in a small pocket on the corner of the fabric. Folded, the Portable Light becomes a shoulder bag, which the Huichol women can carry around during the day as it charges. When the sun goes down, the light, being flexible and weighing less than eight ounces, is easily adapted to different tasks. It can be spread out to serve as a reading mat, draped over the shoulders as a poncho, or rolled up and used as a flashlight.

As she fine-tuned the Portable Light for the Huichol, Kennedy also discovered something new and useful for American society: "Working in the so-called Third World, not only are we bringing people the benefits of a little power, we're also getting great ideas about how we can translate these technologies to our own countries. . . . The idea that we're going to have a top-down centralized system of lighting in our housing and architecture is an historically outdated idea." She and her partner, Frano Violich, have also designed the Soft House, named after Amory Lovins's idea of a soft path for energy — that is, diverse, local, renewable energy sources that match the scale and needs of the consumer. In this house, curtains, flexible walls, and translucent textile screens not only glow with light but also harvest solar energy. Although the Soft House is still in the experimental stage, Kennedy sees it as the path to the future:

> Instead of a centralized grid, imagine a distributed energy network that is literally soft — a flexible network made of multiple, adaptable and co-operative light-emitting textiles that can be touched, held and used by homeowners according to their

needs. . . . The 'soft house' demonstrates the daily experiences of living with textiles that generate power and emit light. Translucent movable curtains along the . . . perimeter convert sunlight into energy throughout the day, shading the house in summer and creating an insulating air layer in winter. Folded downward, a central curtain establishes a habitable off-the-grid energy harvesting room. Folded upward, this luminous curtain becomes a suspended soft chandelier.

Not only do we need to imagine such solutions across cultures and across the globe, but we also need to think back to the past, to ask ourselves whether we are hampered more by brilliance than our ancestors ever were by the dark. It's not too much to imagine that a new night carved out of abundance might also be a time of great possibilities, when we might ask in our way, as Cyril of Jerusalem once did, "What [is] more helpful to wisdom than the night?" And it's not too much to imagine a night with room for more than mere brilliance will allow: the flowering of cockleburs and the warmth of cafés in evening; the safe passage of loggerhead turtles and skyscrapers figured anew; the stars above "more brilliant, more sparklingly gemlike . . . opals you might call them, emeralds, lapis lazuli, rubies, sapphires" and our own long-storied selves intimately at home in immensity.

EPILOGUE

Lascaux Revisited

THE LASCAUX CAVE REMAINS closed to the public, but in the years since Mario Ruspoli created his cinematographic record of the drawings there, the French Ministry of Culture and Communication has built a replica of Lascaux for visitors, and the walls of the cave have been photographed in brilliant color. Archaeologists have examined the drawings with more precise instruments and lenses and now count 1,963 different representations: 915 can be discerned as animals, 434 are signs, and 613 can't be named. There is the one man. They now believe that the horses, with their heavy coats, mark the end of winter and beginning of spring, that the aurochs mark high summer, and that the stags — antlered, depicted in herds — are painted as they were in autumn, just before mating season.

Although the Culture Ministry upgraded the air-conditioning system, in 2001 a technician found mold in the air locks of the entrance site, and within a few weeks the cave floors and ledges were covered in white. Workers suppressed the outbreak with quicklime, but over the next two years, mold continued to grow throughout the cave. In 2003 the ministry began a more comprehensive eradication program, which suppressed the mold once again. Although technicians constantly survey and maintain the site, behind the sealed door of the entrance, the marks of the Paleolithic painters are becoming less and less discernible, and the pigments on the hides of the animals, drawn from memory more than eighteen thousand years ago, are fading.

Meanwhile, our lights draw their own patterns in the dark, as they shine and reflect upward through smoke and ash and cross the same turbulent night winds that make the stars appear to twinkle. If you gaze at the map of earth at night as seen from space, you might imagine the way we appear to astronauts orbiting in the intergalactic silence: come close of day, the earth appears as solids built of illumination and voids created by its absence; in patterns drawn by global drifts of excess and scarcity, thought and afterthought, fortune, innovation, insistence, and accident — patterns that have been accreting for twenty thousand years and conjure no simple feeling. Look once, and you might be amazed at the gift of so much light. Look again, and you might feel sobered by the enormous extent and reach of it. Look yet again, and the countless lights seem to take on unwitting shapes: see the way the crowded headlands of the eastern seaboard make the shape of a head with an outstretched neck, the peninsula of Florida the forelegs, and the Pacific Coast the agile back legs of a fleet stag gathering speed as it rushes headlong into the black Atlantic.

ACKNOWLEDGMENTS

I'm especially grateful to the MacDowell Colony for providing me with the best of places to work, and to the John Simon Guggenheim Memorial Foundation for a fellowship that granted me time to complete this book. The Bowdoin College Library; the Curtis Memorial Library in Brunswick, Maine; and the Maine Interlibrary Loan Service were of immeasurable help to me during my years of research. Gratitude also to the Franklin D. Roosevelt Presidential Library and Museum in Hyde Park, New York; to the New Bedford Whaling Museum Research Library; and to David Low at Consolidated Edison.

I've had the support of many friends during the writing of this book, in particular: Elizabeth Brown, who first put the idea for it in my head; E. F. Weisslitz, who had no end of enthusiasm for it; Andrea Sulzer, always curious; and John Bisbee, who lent me his finely tuned ear for the entire time. Many thanks to Cynthia Cannell for her enduring support on behalf of my work; to Barbara Jatkola for her careful work copyediting the manuscript; and, as always, to Deanne Urmy for her intuition, her precision, her faith.

BIBLIOGRAPHIC NOTE

I am especially indebted to the following books for insight and inspiration: Gaston Bachelard, *The Flame of a Candle*, translated by Joni Caldwell (Dallas: Dallas Institute Publications, 1988); William T. O'Dea, *The Social History of Lighting* (London: Routledge & Kegan Paul, 1958); Wolfgang Schivelbush, *Disenchanted Night: The Industrialization of Light in the Nineteenth Century*, translated by Angela Davies (Berkeley: University of California Press, 1995); and Mario Ruspoli, *The Cave of Lascaux: The Final Photographs* (New York: Harry N. Abrams, 1987).

The first section of *Brilliant* owes much to A. Roger Ekirch, *At Day's Close: Night in Times Past* (New York: W. W. Norton, 2005); Yi-Fu Tuan, "The City: Its Distance from Nature," *Geographical Review* 68, no. 1 (January 1978); Louis-Sébastien Mercier, *Panorama of Paris*, edited by Jeremy D. Popkin (University Park: Pennsylvania University Press, 1999); Richard Ellis, *Men and Whales* (New York: Alfred A. Knopf, 1991); and D. Alan Stevenson, *The World's Lighthouses Before 1820* (London: Oxford University Press, 1959). Chapter 2 owes a particular debt to Wolfgang Schivelbush for insight concerning lanterns and the French Revolution, and to

Yi-Fu Tuan for thoughts on cities and their separation from the natural world.

For the chapters on electricity, I'm grateful to Brian Bowers, *Lengthening the Day: A History of Lighting Technology* (Oxford: Oxford University Press, 1998); Philip Dray, *Stealing God's Thunder: Benjamin Franklin's Lightning Rod and the Invention of America* (New York: Random House, 2005); Jill Jonnes, *Empires of Light: Edison, Tesla, Westinghouse, and the Race to Electrify the World* (New York: Random House, 2004); Robert Friedel and Paul Israel, *Edison's Electric Light: Biography of an Invention* (New Brunswick, NJ: Rutgers University Press, 1987); and Pierre Berton, *Niagara: A History of the Falls* (New York: Kodansha International, 1997).

For the chapters on early-twentieth-century light, I relied largely on Morris Llewellyn Cooke, ed., *Giant Power: Large Scale Electrical Development as a Social Factor* (Philadelphia: Academy of Political and Social Science, 1925); David E. Nye, *Electrifying America: Social Meanings of a New Technology, 1880–1940* (Cambridge, MA: MIT Press, 1992); Katherine Jellison, *Entitled to Power: Farm Women and Technology, 1913–1963* (Chapel Hill: University of North Carolina Press, 1993); Robert A. Caro, *The Years of Lyndon Johnson: The Path to Power* (New York: Alfred A. Knopf, 1982); Mary Ellen Romeo, *Darkness to Daylight: An Oral History of Rural Electrification in Pennsylvania and New Jersey* (Harrisburg: Pennsylvania Rural Electric Association, 1986); James Agee and Walker Evans, *Let Us Now Praise Famous Men: Three Tenant Families* (Boston: Houghton Mifflin, 1988); Barbara Ehrenreich and Deirdre English, *For Her Own Good: 150 Years of the Experts' Advice to Women* (Garden City, NY: Anchor Press, 1978); and Michael J. McDonald and John Muldowny, *TVA and the Dispossessed: The Resettlement of Population in the Norris Dam Area* (Knoxville: University of Tennessee Press, 1982). Chapter 15, especially the section on the sounds of war, owes much to Angus Calder, *The People's War: Britain, 1939–45* (New York: Pantheon Books, 1969).

And for the final section of the book, I'm indebted to A. M. Rosenthal, ed., *The Night the Lights Went Out* (New York: New American Library, 1965); Catherine Rich and Travis Longcore, eds., *Ecological Consequences of Artificial Night Lighting* (Washington, DC: Island Press, 2006); and the International Dark-Sky Association website, http://www.darksky.org.

NOTES

PROLOGUE: THE EARTH AT NIGHT
AS SEEN FROM SPACE

1 "one could not have put": Anton Chekhov, "Easter Eve," in *The Bishop and Other Stories*, trans. Constance Garnett (New York: Ecco Press, 1985), p. 49.
   On a map of the earth: To view the map, see John Weier, "Bright Lights, Big City," http://earthobservatory.nasa.gov/Study/Lights. See also http://visibleearth.nasa.gov (both accessed April 5, 2007).
2 "We are almost certain": Gaston Bachelard, *The Psychoanalysis of Fire*, trans. Alan C. Ross (Boston: Beacon Press, 1968), p. 55.

PART I

5 "Of time that passes": Gaston Bachelard, *The Flame of a Candle*, trans. Joni Caldwell (Dallas: Dallas Institute Publications, 1988), p. 69.

CHAPTER 1: LASCAUX: THE FIRST LAMP

8 In the chambers of Lascaux: The names of the chambers of the Lascaux Cave and the figures in them are from Norbert Aujoulat, *Lascaux: Movement, Space, and Time*, trans. Martin Street (New York: Harry N. Abrams, 2005), p. 30.
9 "The iconography": Ibid., p. 194.
   "Achieving full and accurate": Sophie A. de Beaune and Randall White, "Ice Age Lamps," *Scientific American*, March 1993, p. 112.
10 "render to God": *Asser's Life of King Alfred*, trans. L. C. Jane (New York: Cooper Square, 1966), pp. 85–87.

11 "an object like the ghost": Charles Dickens, *Great Expectations* (Boston: Bedford Books, 1996), p. 337.

12 "It was said that": Alice Morse Earle, *Home Life in Colonial Days* (Stockbridge, MA: Berkshire House, 1993), p. 34.

"a serious undertaking": Harriet Beecher Stowe, *Poganuc People: Their Lives and Loves* (New York: Fords, Howard & Hulbert, 1878), p. 230.

13 "cut very small": Arthur H. Hayward, *Colonial Lighting* (New York: Dover Publications, 1962), pp. 84–85.

"even the best-read people": Marshall B. Davidson, "Early American Lighting," *Metropolitan Museum of Art Bulletin*, n.s., 3, no. 1 (Summer 1944): 30.

14 "There are several Ways": Jonathan Swift, "Directions to Servants," *Directions to Servants and Miscellaneous Pieces, 1733–1742*, ed. Herbert Davis (Oxford: Basil Blackwell, 1959), pp. 14–15.

"stinking tallow": William Shakespeare, *Cymbeline*, in *The Riverside Shakespeare* (Boston: Houghton Mifflin, 1974), p. 1529.

15 "At the Court": William T. O'Dea, *The Social History of Lighting* (London: Routledge & Kegan Paul, 1958), p. 37.

"In the middle": Jean Verdon, *Night in the Middle Ages*, trans. George Holoch (Notre Dame, IN: University of Notre Dame Press, 2002), p. 77.

"Their fire sticks": Dr. A. S. Gatschet, quoted in Walter Hough, *Fire as an Agent in Human Culture*, Smithsonian Institution Bulletin, no. 139 (Washington, DC: Government Printing Office, 1926), p. 99.

16 "a cold dark frosty": *The Tinder Box* (London: William Marsh, 1832), quoted in O'Dea, *The Social History of Lighting*, p. 237.

"About two o'clock": James Boswell, quoted in Molly Harrison, *The Kitchen in History* (New York: Charles Scribner's Sons, 1972), pp. 92–93.

"unfortunate man staying": Jane C. Nylander, *Our Own Snug Fireside: Images of the New England Home, 1760–1860* (New Haven, CT: Yale University Press, 1994), p. 107.

17 "The English dwell": Quoted in A. Roger Ekirch, *At Day's Close: Night in Times Past* (New York: W. W. Norton, 2005), p. 48.

"found it a matter": John Smeaton, quoted in O'Dea, *The Social History of Lighting*, p. 224.

"A French *Book of Trades*": Ekirch, *At Day's Close*, p. 156.

"From Easter to Saint-Rémi": Verdon, *Night in the Middle Ages*, p. 111.

18 "A servant would have": Cyril of Jerusalem, in Philip Schaff and
    Henry Wace, eds., *A Select Library of Nicene and Post-Nicene Fathers of
    the Christian Church*, 2nd. ser., 7 (New York: Christian Literature,
    1894), p. 52.
    "And what [is] more": Ibid., pp. 52–53.
    "in orderly rows": Gertrude Whiting, *Tools and Toys of Stitchery* (New
    York: Columbia University Press, 1928), p. 253.

CHAPTER 2: TIME OF DARK STREETS

20 "The light of the sun": Libanius, quoted in M. Luckiesh, *Artificial
    Light: Its Influence upon Civilization* (New York: Century, 1920),
    p. 153.
    "Hang-chou boasted": Yi-Fu Tuan, "The City: Its Distance from
    Nature," *Geographical Review* 68, no. 1 (January 1978): 9.
    "No oil lamps lighted": Jérôme Carcopino, *Daily Life in Ancient Rome:
    The People and the City at the Height of the Empire*, ed. Henry T. Row-
    ell (New Haven, CT: Yale University Press, 1940), p. 47.

22 "About half a league": Jean-Jacques Rousseau, quoted in A. Roger
    Ekirch, *At Day's Close: Night in Times Past* (New York: W. W. Norton,
    2005), p. 63.
    "as if it were in tyme": Fynes Moryson, quoted ibid., p. 61.
    "maintained more than": Ekirch, *At Day's Close*, p. 64.
    "At night all houses": Quoted in Wolfgang Schivelbush, *Disenchanted
    Night: The Industrialization of Light in the Nineteenth Century*, trans.
    Angela Davies (Berkeley: University of California Press, 1995),
    p. 81.

23 "whose feet in many towns": Jean Verdon, *Night in the Middle Ages*,
    trans. George Holoch (Notre Dame, IN: University of Notre Dame
    Press, 2002), p. 85.

24 "no man [may] walke": Quoted in G. T. Salusbury-Jones, *Street Life
    in Medieval England* (Sussex, Eng.: Harvester Press, 1975), p. 139.
    "Let no one be so bold": Quoted in Verdon, *Night in the Middle Ages*,
    p. 80.

25 "It has been said": Luckiesh, *Artificial Light*, p. 153.

26 "On the twenty-sixth day": Quoted in Verdon, *Night in the Middle
    Ages*, p. 124.
    "A man would thincke": Quoted in Ekirch, *At Day's Close*, p. 71.

27 "a lamp that *waits*": Gaston Bachelard, *The Flame of a Candle*, trans.

Joni Caldwell (Dallas: Dallas Institute Publications, 1988), pp. 71–72.

"On 1 December": Quoted in Schivelbush, *Disenchanted Night*, pp. 90–91.

"the magistrates": Edwin G. Burrows and Mike Wallace, *Gotham: A History of New York City to 1898* (New York: Oxford University Press, 1999), p. 111.

28 "totally inadequate to dispel": William Sidney, *England and the English in the Eighteenth Century: Chapters in the Social History of the Times*, vol. 1 (London: Ward & Downey, 1892), p. 15.

29 "The light, such as it was": Ibid., pp 14–15.

"greasy clodhopping fellows": Ibid., p. 15.

"Another thing; they might": Louis-Sébastien Mercier, *Panorama of Paris*, ed. Jeremy D. Popkin (University Park: Pennsylvania University Press, 1999), p. 43.

"Cautious citizens in Birmingham": Tuan, "The City," p. 10.

"that as the fear": Ibid.

30 "in Vienna in 1688": Craig Koslofsky, "Court Culture and Street Lighting in Seventeenth-Century Europe," *Journal of Urban History* 28, no. 6 (September 2002): 760.

"the streets after ten": Mercier, *Panorama of Paris*, p. 132.

31 "was an undertaking": Sidney, *England and the English*, p. 15.

"a custom, both in ancient": Leone di Somi, *Dialogues on Stage Affairs*, quoted in Frederick Penzel, *Theatre Lighting Before Electricity* (Middletown, CT: Wesleyan University Press, 1978), p. 7.

32 "Until he himself": William J. Lawrence, *Old Theatre Days and Ways* (New York: Benjamin Bloom, 1968), p. 130.

"These beautiful lights": Johannes Neiner, quoted in Koslofsky, "Court Culture and Street Lighting," p. 751.

"Night falls": Mercier, *Panorama of Paris*, p. 95.

33 "In the old days": Ibid., p. 41.

"I have known fogs": Ibid., pp. 133–34.

"The darkness that spread": Schivelbush, *Disenchanted Night*, p. 106.

34 "the clanking of its huge axe": Thomas Carlyle, *The French Revolution: A History* (New York: Modern Library, n.d.), p. 625.

"In the summer of 1789": Schivelbush, *Disenchanted Night*, p. 100.

"Originally, this word": Mercier, quoted ibid.

"the gaunt scarecrows": Charles Dickens, *A Tale of Two Cities* (New York: Signet, 1997), p. 39.

"not infrequently, the hapless": Quoted in Schivelbush, *Disenchanted Night*, p. 100n.

"whirled across the Place": Carlyle, *The French Revolution*, p. 164.

35 "the order of nature": Philip Balthasar Sinold, quoted in Koslofsky, "Court Culture and Street Lighting," p. 746.

"now [opened] hardly": Friedrich Justin Bertuch, quoted in Koslofsky, "Court Culture and Street Lighting," p. 744.

36 "The city lives": Richard Eder, "New York," in "Cities in Winter," *Saturday Review*, January 8, 1977, p. 25.

"not a small New York": Elizabeth Hardwick, "Boston," in *A View of My Own: Essays in Literature and Society* (London: William Heinemann, 1964), p. 151.

CHAPTER 3: LANTERNS AT SEA

37 "more scarce than": Herman Melville, *Moby Dick* (New York: Penguin Books, 1992), p. 466.

"The oil is hissing": J. Ross Browne, quoted in Richard Ellis, *Men and Whales* (New York: Alfred A. Knopf, 1991), p. 198.

38 "When the flesh": From Arrian's description of the conquests of Alexander the Great, quoted ibid., p. 33.

39 "When they come within": Levi Whitman, quoted in James Deetz and Patricia Scott Deetz, *The Times of Their Lives: Life, Love, and Death in Plymouth Colony* (New York: Anchor Books, 2001), p. 248.

"The respiratory canal": William Davis, *Nimrod of the Sea*, quoted in Alexander Starbuck, *History of the American Whale Fishery* (Secaucus, NJ: Castle Books, 1989), p. 157.

"carpet a room": Ibid., p. 156.

40 "The lips and throat": Ibid., p. 157.

"subsists wholly on mist": *The King's Mirror*, trans. Laurence Marcellus Larson (New York: American-Scandinavian Foundation, 1917), p. 123.

"that wondrous Venetian blind": Melville, *Moby Dick*, p. 297.

"It is as if": Ibid., p. 461.

41 "the unmelted skin": Ellis, *Men and Whales*, p. 198.

"like the left wing": Melville, *Moby Dick*, p. 462.

42 "'Bible leaves!'": Ibid., p. 460.

"There they lay": Ibid., p. 466.

43 "a new kind of Candles": *The Papers of Benjamin Franklin*, quoted in

Richard C. Kugler, *The Whale Oil Trade, 1750–1775* (New Bedford, MA: Old Dartmouth Historical Society, 1980), p. 13n.

44 "In the great Sperm Whale": Melville, *Moby Dick*, p. 379.

46 "whether Leviathan": Ibid., p. 501.

47 "They think that at best": Ibid., pp. 118–19.

48 "The only danger": Pliny the Elder, *The Natural History of Pliny*, trans. John Bostock and H. T. Riley, vol. 6 (London: Henry G. Bohn, 1858), p. 339.

49 "was to drive": D. Alan Stevenson, *The World's Lighthouses Before 1820* (London: Oxford University Press, 1959), p. xxiv.
"Many coastal villages": Bella Bathurst, *The Lighthouse Stevensons: The Extraordinary Story of the Building of the Scottish Lighthouses by the Ancestors of Robert Louis Stevenson* (New York: HarperCollins, 1999), p. 26.
"The rust-colored gneiss": Ibid., p. 54.

50 "At midsummer the party": Stevenson, *The World's Lighthouses*, p. 115.

51 "Quickly the fire": Ibid., p. 121.
"Fenders fixed": Ibid., p. 124.

52 "very strong and bright": John Smeaton, quoted ibid., pp. 125–26.

53 "So long as the air": Samuel Williams, quoted in *Harvard Case Histories in Experimental Science*, ed. James Bryant Conant, case 2, *The Overthrow of the Phlogiston Theory: The Chemical Revolution of 1775–1789* (Cambridge, MA: Harvard University Press, 1964), p. 15.
"As soon as the Air": Ibid., pp. 15–16.

54 "very white": Quoted in Brian Bowers, *Lengthening the Day: A History of Lighting Technology* (Oxford: Oxford University Press, 1998), p. 28.
"as the light emitted": A.F.M. Willich, *The Domestic Encyclopaedia, or A Dictionary of Facts, and Useful Knowledge*, vol. 3 (London: B. McMillan, 1802), s.v. "lamp," http://chestofbooks.com/reference/The-Domestic-Encyclopaedia-Vol3/Lamp.html (accessed June 29, 2009).

55 "Being 'the thing'": Marshall B. Davidson, "Early American Lighting," *Metropolitan Museum of Art Bulletin*, n.s., 3, no. 1 (Summer 1944): 37.
"The modest versions": Ibid.
"the single most powerful": Stevenson, *The World's Lighthouses*, p. xix.

56 "Every night they go": Henry Beston, *The Outermost House: A Year of*

Life on the Great Beach of Cape Cod (New York: Henry Holt, 1992), p. 128.

57 "There has just been": Ibid., pp. 116–17, 121.

CHAPTER 4: GASLIGHT

59 "It seldom needs": Thomas Cooper, *Some Information Concerning Gas Lights* (Philadelphia: John Conrad, 1816), p. 23.

"The inflammable gas": Philippe Lebon, quoted in Wolfgang Schivelbush, *Disenchanted Night: The Industrialization of Light in the Nineteenth Century*, trans. Angela Davies (Berkeley: University of California Press, 1995), p. 23.

60 "All factories": M. E. Falkus, "The Early Development of the British Gas Industry, 1790–1815," *Economic History Review*, n.s., 35, no. 2 (May 1982): 219.

61 "It was estimated": Ibid., p. 223.

"Suppose it were required": Cooper, *Some Information Concerning Gas Lights*, p. 12.

62 "The burners were simply": William T. O'Dea, *The Social History of Lighting* (London: Routledge & Kegan Paul, 1958), p. 115.

"Clean and orderly": Quoted in Francis D. Klingender, *Art and the Industrial Revolution* (London: Noel Carrington, 1947), p. 111.

63 "This spire increases": John Buddle, quoted in T. S. Ashton and Joseph Sykes, *The Coal Industry of the Eighteenth Century* (New York: Augustus M. Kelley, 1967), p. 44n.

"Clad from head to foot": Ibid., pp. 44–45.

64 "Everything in the way": Quoted ibid., p. 42n.

"were about three hundred": Quoted ibid., p. 49n.

65 "work was continued": T. E. Forster, "Historical Notes on Wallsend Colliery," *Transactions of the Institution of Mining Engineers* 15 (1897–1898), http://www.dmm-gallery.org.uk/transime/u15f-01.htm (accessed February 1, 2009).

"sometimes tried to carry on": Ashton and Sykes, *The Coal Industry*, p. 51.

66 "had provided the miner": Ibid., p. 53.

"if it were intended": Sir Humphry Davy, quoted in Samuel Clegg Jr., *Practical Treatise on the Manufacture and Distribution of Coal-Gas* (London: John Weale, 1841), p. 17.

"Winsor was not": Schivelbush, *Disenchanted Night*, pp. 26–27.

67 "a brightness clear": Quoted in Clegg, *Practical Treatise*, pp. 20–21.

"I foresee in this": Charles Dickens, *The Lamplighter: A Farce* (London: Printed from a Manuscript in the Forster Collection at the South Kensington Museum, 1879), p. 10.

68 "It was strangely believed": Clegg, *Practical Treatise*, p. 17.

69 "Wherever a gas-factory": Quoted in Lynda Nead, *Victorian Babylon: People, Streets and Images in Nineteenth-Century London* (New Haven, CT: Yale University Press, 2000), p. 94.

"Mr. Arabin, deposed": Cooper, *Some Information Concerning Gas Lights*, p. 131.

70 "When the effluvia": Ibid., p. 133.

"*Thomas Edgely* is": Ibid., pp. 134–35.

"at present it is": Quoted in Schivelbush, *Disenchanted Night*, p. 35.

71 "In 1821 no town": Steven J. Goldfarb, "A Regency Gas Burner," *Technology and Culture* 12, no. 3 (July 1971): 476.

"Paris was illuminated": Quoted in Walter Benjamin, *The Arcades Project*, trans. Howard Eiland and Kevin McLaughlin (Cambridge, MA: Belknap Press of Harvard University Press, 1999), p. 565.

"The work of Prometheus": Robert Louis Stevenson, "A Plea for Gas Lamps," in *Virginibus Puerisque and Other Papers* (New York: Charles Scribner's Sons, 1893), p. 274.

72 "Paris will be": Vincent van Gogh to Theo van Gogh, letter 550, in *The Complete Letters of Vincent van Gogh*, vol. 3 (Greenwich, CT: New York Graphic Society, 1959), p. 75.

"The whole of Paris": Andreas Bluhm and Louise Lippincott, *Light! The Industrial Age, 1750–1900* (New York: Thames & Hudson, 2001), p. 182.

73 "During the day": Karl Gutzkow, quoted in Benjamin, *The Arcades Project*, p. 537.

74 "The new mode of illumination": Frederick Penzel, *Theatre Lighting Before Electricity* (Middletown, CT: Wesleyan University Press, 1978), p. 54.

"a *kaleidoscope*": Charles Baudelaire, quoted in Walter Benjamin, "On Some Motifs in Baudelaire," in *Illuminations: Essays and Reflections*, ed. Hannah Arendt, trans. Harry Zohn (New York: Schocken Books, 1969), p. 175.

"As the darkness came on": Edgar Allan Poe, "The Man of the

Crowd," in *The Unabridged Edgar Allan Poe* (Philadelphia: Running Press, 1983), p. 648.

75 "As the night deepened": Ibid., p. 650.

76 "Some rushed about": "Bereft of Light: Terrific Explosion at the Metropolitan Gas Works," *New York Times*, December 24, 1871, p. 5.

CHAPTER 5: TOWARD A MORE PERFECT FLAME

77 "A candle, you know": Michael Faraday, *The Chemical History of a Candle* (Mineola, NY: Dover Publications, 2002), p. 13.

78 "brilliantly white, inodorous": Campbell Morfit, *A Treatise on Chemistry Applied to the Manufacture of Soap and Candles* (Philadelphia: Parry & McMillan, 1856), p. 543.
"mortal man should feed": Herman Melville, *Moby Dick* (New York: Penguin Books, 1992), p. 325.

79 "any common use": "Camphene and Burning Fluid," *New York Times*, November 28, 1854, p. 4.
"a burning fluid lamp": Jane Nylander, "Two Brass Lamps . . . ," *Historic New England Magazine*, Winter/Spring 2003, http://www.historicnewengland.org/nehm/2003winterspringpage04.htm (accessed February 12, 2009).

80 "If possible avoid": Quoted in Charles Panati, *Panati's Extraordinary Origins of Everyday Things* (New York: Harper & Row, 1989), p. 109.
"The chemical match": Quoted in Walter Benjamin, *The Arcades Project*, trans. Howard Eiland and Kevin McLaughlin (Cambridge, MA: Belknap Press of Harvard University Press, 1999), p. 568.

81 "We dreamed of the lamp": Gaston Bachelard, *The Flame of a Candle*, trans. Joni Caldwell (Dallas: Dallas Institute Publications, 1988), p. 66.

83 "could supply a family's": Daniel Yergin, *The Prize: The Epic Quest for Oil, Money, and Power* (New York: Simon & Schuster, 1992), p. 34.
"the kind of oil": William T. O'Dea, *The Social History of Lighting* (London: Routledge & Kegan Paul, 1958), pp. 55–56.
"The production of petroleum": *Titusville Morning Herald*, quoted in Harold F. Williamson and Arnold R. Daum, *The American Petroleum Industry: The Age of Illumination, 1859–1899* (Evanston, IL: Northwestern University Press, 1959), p. 371.

85 "The country has been flooded": Quoted in Kathleen Grier, *The Popular Illuminator: Domestic Lighting in the Kerosene Era, 1860–1900* (Rochester, NY: Strong Museum, 1985), p. 10.

"Good oil poured": Catharine E. Beecher and Harriet Beecher Stowe, *The American Woman's Home* (1869; repr., Whitefish, MT: Kessinger Publishing, 2004), p. 190.

86 "Wed., September 1": *Willimantic Chronicle,* quoted in "The Dangers of Kerosene Lamps," http://www.thelampworks.com/lw_lamp_accidents.htm (accessed June 3, 2009).

87 "There is as much wit": *The Woman's Book,* vol. 2 (New York: Charles Scribner's Sons, 1894), quoted in Grier, *The Popular Illuminator,* pp. 7–8.

88 "By keeping their independent": Wolfgang Schivelbush, *Disenchanted Night: The Industrialization of Light in the Nineteenth Century,* trans. Angela Davies (Berkeley: University of California Press, 1995), p. 162.

"I boldly declare": Quoted in Benjamin, *The Arcades Project,* p. 562.

89 "It seems there are": Bachelard, *The Flame of a Candle,* p. 4.

PART II

91 "You turn the thumbscrew": *New York Times,* September 5, 1882, p. 8.

CHAPTER 6: LIFE ELECTRIC

93 "little click": Gaston Bachelard, *The Flame of a Candle,* trans. Joni Caldwell (Dallas: Dallas Institute Publications, 1988), p. 64.

"it seemed to live": Park Benjamin, *The Age of Electricity from Amber-Soul to Telephone* (New York: Charles Scribner's Sons, 1888), pp. 2–3.

95 "He suspended a long": Quoted ibid., p. 11.

96 "When a nail": Ewald von Kleist, quoted ibid., p. 15.

97 "to move electricity": Philip Dray, *Stealing God's Thunder: Benjamin Franklin's Lightning Rod and the Invention of America* (New York: Random House, 2005), p. 49.

"I advise you": Quoted in Jill Jonnes, *Empires of Light: Edison, Tesla, Westinghouse, and the Race to Electrify the World* (New York: Random House, 2004), p. 23.

"a vast country": Albrecht von Haller, quoted in Dray, *Stealing God's Thunder*, p. 46.

98 "chagrined a little": Benjamin Franklin, "The Electrical Writings of Benjamin Franklin and Friends," collected by Robert A. Morse, 2004, Wright Center for Innovation in Science Teaching, Tufts University, Medford, MA, p. 24 (pdf p. 35), http://www.tufts.edu/as /wright_center/personal_pages/bob_m/franklin_electricity_screen. pdf (accessed June 29, 2009).

"I have lately made": Ibid., p. 58 (pdf p. 69).

"was the first to discover": Dray, *Stealing God's Thunder*, pp. 54–55.

99 "There was scarce": Quoted ibid., p. 67.

"There is something": Franklin, quoted ibid., p. 57.

"be kept clean": Franklin, "The Electrical Writings," p. 45 (pdf p. 56).

100 "As soon as any": Ibid., p. 95 (pdf p. 106).

"Franklin's conclusions demanded": Dray, *Stealing God's Thunder*, p. 83.

101 "I obtain several": Alessandro Volta, quoted in Edwin J. Houston, *Electricity in Every-Day Life*, vol. 1 (New York: P. F. Collier & Son, 1905), pp. 347–49.

"made electricity manageable": Benjamin, *The Age of Electricity*, p. 32.

103 "the city of Moscow": Francis R. Upton, "Edison's Electric Light," *Scribner's*, February 1880, p. 532.

104 "suddenly found themselves": Quoted in Wolfgang Schivelbush, *Disenchanted Night: The Industrialization of Light in the Nineteenth Century*, trans. Angela Davies (Berkeley: University of California Press, 1995), p. 55.

105 "A new sort of urban star": Robert Louis Stevenson, "A Plea for Gas Lamps," in *Virginibus Puerisque and Other Papers* (New York: Charles Scribner's Sons, 1893), pp. 277–78.

"Promptly as the courthouse": Quoted in John Winthrop Hammond, *Men and Volts: The Story of General Electric* (New York: J. B. Lippincott, 1941), pp. 31–32.

106 "The city's 65 gas lamps": Richard B. Biever, "Indiana's Bright Lights," *Electric Consumer*, Indiana Statewide Association of Rural Electric Cooperatives, http://indremcs.org/ec/article (accessed February 13, 2009).

"lighting . . . emerged": David E. Nye, *Electrifying America: Social Meanings of a New Technology, 1880–1940* (Cambridge, MA: MIT Press, 1992), p. 54.

"Cities lit in this way": Schivelbush, *Disenchanted Night*, p. 126.

107 "following the practice": Howard Strong, "The Street Beautiful in Minneapolis," in *American City*, vol. 9 (New York: Civic Press, 1913), pp. 228–29.

108 "single light alone": "Lights for a Great City: Brush's System in Successful Use Last Night," *New York Times*, December 21, 1880, p. 2.

"the moment the dazzling": Ibid.

109 "Since the electric lamps": "Lights in Street Lamps: Bright Electricity Makes the Old-Time Gas Look Dim," *New York Times*, June 20, 1898, p. 10.

CHAPTER 7: INCANDESCENCE

111 "described an experiment": Quoted in Brian Bowers, *Lengthening the Day: A History of Lighting Technology* (Oxford: Oxford University Press, 1998), p. 89.

"It was all before me": Thomas Edison, quoted in Paul Israel, *Edison: A Life of Invention* (New York: John Wiley & Sons, 1998), p. 166.

112 "ran from the instrument": *New York Sun*, quoted in Israel, *Edison*, p. 165.

"Now that I have": Edison, quoted in George Westinghouse, "A Reply to Mr. Edison," *North American Review*, December 1889, p. 655.

"A mistaken idea": Francis R. Upton, "Franklin's Electric Light," *Scribner's*, February 1880, p. 531.

114 "When I was a boy": David Trumbull Marshall, *Recollections of Boyhood Days in Old Metuchen* (Flushing, NY: Case Publishing, 1930), in Metuchen Edison History Features, http://www.jhalpin.com/metuchen/history/boy37.htm (accessed January 18, 2006).

"His iron ideas": *New York Daily Graphic*, quoted in Jill Jonnes, *Empires of Light: Edison, Tesla, Westinghouse, and the Race to Electrify the World* (New York: Random House, 2004), p. 54.

115 "At six o'clock": *New York Herald*, quoted in Robert Friedel and Paul Israel, *Edison's Electric Light: Biography of an Invention* (New Brunswick, NJ: Rutgers University Press, 1987), p. 37.

"The more resistance": Edison, quoted in Friedel and Israel, *Edison's Electric Light*, p. 75.

116 "(April 29) Wood loop": Friedel and Israel, *Edison's Electric Light*, p. 154.

117 "Edison's electric light": *New York Herald*, quoted in Jonnes, *Empires of Light*, p. 65.

118 "The light was subjected": *New York Herald*, quoted in Friedel and Israel, *Edison's Electric Light*, pp. 112–13.

119 "Standing at one end": *Lowell Morning Mail*, quoted in David E. Nye, *Electrifying America: Social Meanings of a New Technology, 1880–1940* (Cambridge, MA: MIT Press, 1992), p. 190.

120 "It was a great deal": Herbert L. Satterlee, *J. Pierpont Morgan: An Intimate Portrait* (New York: Macmillan, 1939), p. 207.
"had to be run by": Ibid., p. 208.

121 "All that changed": Jonnes, *Empires of Light*, pp. 79–80.

122 "It was a light": "Miscellaneous City News: Edison's Electric Light," *New York Times*, September 5, 1882, p. 8.

123 "In the stores": *New York Herald*, quoted in Friedel and Israel, *Edison's Electric Light*, p. 222.

124 "I would get a fever": Nikola Tesla, quoted in Pierre Berton, *Niagara: A History of the Falls* (New York: Kodansha International, 1997), pp. 157–58.

125 "Spare me that nonsense": Edison, quoted in Berton, *Niagara*, p. 161.
"It will never be free": Edison, quoted in Margaret Cheney, *Tesla: Man Out of Time* (New York: Dorset Press, 1981), p. 43.
"The wind at times": "In a Blizzard's Grasp," *New York Times*, March 13, 1888, p. 1.
"Poles with their long arms": "Wires Down Everywhere," *New York Times*, March 13, 1888, pp. 1–2.

126 "The man appeared": Quoted in Jill Jonnes, "New York Unplugged, 1889," *New York Times*, August 13, 2004, http://www.nytimes.com (accessed June 28, 2009).
"aspect of the city": "A Night of Darkness: More Than One Thousand Electric Lights Extinguished," *New York Times*, October 15, 1889, p. 2.

127 "As to the accidents": Westinghouse, "A Reply to Mr. Edison," p. 661.

CHAPTER 8: OVERWHELMING
BRILLIANCE: THE WHITE CITY

128 "Electricity is the half": Hubert Howe Bancroft, *The Book of the Fair: An Historical and Descriptive Presentation of the World's Science, Art, and Industry, as Viewed Through the Columbian Exposition at Chicago in 1893* (Chicago: Bancroft, 1893), p. 399.

"a marsh when": Julian Ralph, "Our Exposition at Chicago, with Plan of Exposition Grounds and Buildings," *Harper's*, January 1892, p. 206.

"a treacherous morass": Quoted in Norma Bolotin and Christine Laing, *The World's Columbian Exposition: The Chicago World's Fair of 1893* (Urbana: University of Illinois Press, 2002), p. 11.

129 "of darkened ivory": Ralph, "Our Exposition at Chicago," p. 207.

"so bewildering no eye": W. E. Cameron, quoted in Marc J. Seifer, *Wizard: The Life and Times of Nikola Tesla, Biography of a Genius* (New York: Citadel Press, 1998), p. 117.

"There would be a dozen": Quoted in Bolotin and Laing, *The World's Columbian Exposition*, p. 148.

130 "It is the part": Bancroft, *The Book of the Fair*, p. 401.

"were used lavishly": J. P. Barrett, *Electricity at the Columbian Exposition* (Chicago: R. R. Donnelley & Sons, 1894), p. 1.

131 "brilliance almost too dazzling": Bancroft, *The Book of the Fair*, p. 402.

"Having seen nothing": Quoted in Erik Larson, *The Devil in the White City: Murder, Magic, and Madness at the Fair That Changed America* (New York: Crown, 2003), p. 254.

"by virtue of pressure": Louis H. Sullivan, *The Autobiography of an Idea* (New York: Dover Publications, 1956), p. 308.

132 "Chicago, one might say": William Dean Howells, *Letters of an Altrurian Traveller* (Gainesville, FL: Scholars' Facsimiles & Reprints, 1961), p. 20.

"'Undisciplined' — that is the word": H. G. Wells, "The Future in America: A Search After Realities," *Harper's Weekly*, July 21, 1906, p. 1020.

133 "Within the memory": George Bird Grinnell, *Blackfoot Lodge Tales: The Story of a Prairie People* (Lincoln: University of Nebraska Press, 1970), pp. 200–201.

"The Native Americans": Robert W. Rydell, *All the World's a Fair:*

Visions of Empire at American International Expositions, 1876–1916 (Chicago: University of Chicago Press, 1984), p. 63.

"several of the exhibits": Ibid.

134 "Here was an opportunity," Rossiter Johnson, ed., *A History of the World's Columbian Exposition Held in Chicago in 1893*, vol. 3 (New York: D. Appleton, 1898), pp. 433–34.

"of whom twenty-one": Ibid., p. 444.

"Sight-seers . . . were fascinated": Ibid.

135 "As if to shame": Frederick Douglass, introduction to *The Reason Why the Colored American Is Not in the World's Columbian Exposition*, by Ida B. Wells, Frederick Douglass, Irvine Garland Penn, and Ferdinand L. Barnett, ed. Robert W. Rydell (Urbana: University of Illinois Press, 1999), p. 13.

"When it was ascertained": Ferdinand L. Barnett, "The Reason Why," in *The Reason Why the Colored American Is Not in the World's Columbian Exposition*, pp. 74–75.

"the contents of a great": Quoted in William Cronon, *Nature's Metropolis: Chicago and the Great West* (New York: W. W. Norton, 1991), p. 344.

136 "We earnestly desired": Douglass, introduction, pp. 7, 16.

137 "his gaze turned upward": Bancroft, *The Book of the Fair*, p. 403.

"no two of which": Barrett, *Electricity at the Columbian Exposition*, p. 18.

138 "The Edison tower": Bancroft, *The Book of the Fair*, p. 424.

"Close at hand": Ibid., pp. 421–22.

139 "dipping reels of wire": Ibid., p. 409.

"When the currents": Quoted in Seifer, *Wizard*, p. 121.

140 "lumped off the whole": Quoted ibid., p. 120.

"without injury to life": Quoted ibid.

"The streams of light": Nikola Tesla and Thomas Commerford Martin, *The Inventions, Researches, and Writings of Nikola Tesla: With Special Reference to His Work in Polyphase Current and High Potential Lighting*, 2nd ed. (New York: Electrical Engineer, 1894), p. 320.

"After such a striking,": Quoted in Margaret Cheney, *Tesla: Man Out of Time* (New York: Dorset Press, 1981), p. 73.

141 "alive with the deafening": Jill Jonnes, *Empires of Light: Edison, Tesla, Westinghouse, and the Race to Electrify the World* (New York: Random House, 2004), p. 267.

"Popular interest": Francis E. Leupp, *George Westinghouse: His Life and Achievements* (Boston: Little, Brown, 1919), p. 169.

"What astonished visitors": Ibid.

142 Winslow Homer's *The Fountains:* This painting is in the collection of the Bowdoin College Museum of Art, Brunswick, Maine.

143 "I believe": Quoted in David F. Burg, *Chicago's White City of 1893* (Lexington: University Press of Kentucky, 1976), p. 287.

There are 'bits'": "Fate of the Chicago World's Fair Buildings," *Scientific American*, October 3, 1896, American Periodical Series Online, p. 267.

CHAPTER 9: NIAGARA: LONG-DISTANCE LIGHT

144 "I was in a manner": Charles Dickens, *American Notes for General Circulation*, vol. 2 (London: Chapman & Hall, 1842), pp. 177–78.

145 "All the coal raised": Sir William Siemens, quoted in Pierre Berton, *Niagara: A History of the Falls* (New York: Kodansha International, 1997), p. 51.

"The greatest and strongest": Peter Kalm, quoted in Charles Mason Dow, *Anthology and Bibliography of Niagara Falls*, vol. 1 (Albany: State of New York, 1921), p. 56.

146 "Several of the *French*": Ibid., p. 58.

147 "Thirteen hundred workmen": Berton, *Niagara*, p. 162.

149 "[I] pictured in my imagination": Nikola Tesla, quoted in Marc J. Seifer, *Wizard: The Life and Times of Nikola Tesla, Biography of a Genius* (New York: Citadel Press, 1998), p. 132.

"the inlet gates": Jill Jonnes, *Empires of Light: Edison, Tesla, Westinghouse, and the Race to Electrify the World* (New York: Random House, 2004), p. 320.

"the falls and Buffalo": Tesla, quoted ibid., p. 326.

"Electrical experts say": *Buffalo Enquirer*, quoted in Jonnes, *Empires of Light*, pp. 328–29.

150 "Wherever mankind wishes": Irving Fisher, "The Decentralization and Suburbanization of Population," in *Giant Power: Large Scale Electrical Development as a Social Factor*, ed. Morris Llewellyn Cooke (Philadelphia: American Academy of Political and Social Science, 1925), p. 96.

"Yoked to the Cataract!": *Buffalo Enquirer,* quoted in Jonnes, *Empires of Light,* p. 329.

"What *is* electricity": R. R. Bowker, ed., "Electricity," no. 12 in The Great American Industries series, *Harper's,* October 1896, p. 710.

"Now, I must tell you": Tesla, quoted in Seifer, *Wizard,* p. 5.

151   "To Adams the dynamo": Henry Adams, *The Education of Henry Adams: An Autobiography* (Boston: Houghton Mifflin, 1918), p. 380.

152   "The dynamos and turbines": H. G. Wells, "The Future in America: A Search After Realities," *Harper's Weekly,* July 21 1906, p. 1019.

PART III

153   "So if we moderns": Fernand Braudel, *Capitalism and Material Life, 1400–1800,* trans. Miriam Kochan (New York: Harper & Row, 1973), p. 226.

CHAPTER 10: NEW CENTURY, LAST FLAME

155   "In our households": Edwin J. Houston, *Electricity in Every-Day Life,* vol. 1 (New York: P. F. Collier & Son, 1905), p. 1.

Gas, for instance: Information on the cost of lighting comes from M. Luckiesh, *Artificial Light: Its Influence upon Civilization* (New York: Century, 1920), pp. 214–17.

156   "Houses were lit": Richard K. Nelson, *Make Prayers to the Raven: A Koyukon View of the Northern Forest* (Chicago: University of Chicago Press, 1983), p. 18.

157   "the constellation of": Walter Hough, "The Lamp of the Eskimo," in *The Annual Report of the Board of Regents of the Smithsonian Institution Showing the Operations, Expenditures, and the Condition of the Institution for the Year Ending June 30, 1896: Report of the U.S. National Museum* (Washington, DC: Government Printing Office, 1898), p. 1038.

158   "Had he smoked": *Ward's Auto World,* October 1970, p. 63, quoted in "Lamp Fillers: Notes and Queries, Quotes and News: Lamp Pollution?" *History of Lamps and Lighting: The Rushlight Archives, 1934–2006,* DVD, Rushlight Club, 2007.

159   "Lamp trimming only reaches": Hough, "The Lamp of the Eskimo," p. 1034.

"The Eskimo have": Walter Hough, "The Origin and Range of the Eskimo Lamp," *American Anthropologist* 11, no. 4 (April 1898): 117.

"unlike our lighting systems": Walter Benjamin, "The Lamp," in *Selected Writings*, vol. 2, *1927–1934*, ed. Michael W. Jennings, Howard Eiland, and Gary Smith, trans. Rodney Livingstone and others (Cambridge, MA: Belknap Press of Harvard University Press, 1999), p. 692.

CHAPTER 11: GLEAMING THINGS

161 "In days of old": Edward Hungerford, "Night Glow of the City," *Harper's Weekly*, April 30, 1910, p. 13.

162 "when it was found": "Fines the Edison Co. for Smoke Nuisance," *New York Times*, January 17, 1911, p. 7.

164 "Electrical articles": Quoted in Ronald C. Tobey, *Technology as Freedom: The New Deal and the Electrical Modernization of the American Home* (Berkeley: University of California Press, 1996), p. 30.

"It was a Dover iron": Quoted in Earl Lifshey, *The Housewares Story: A History of the American Housewares Industry* (Chicago: National Housewares Manufacturers Association, 1973), p. 231.

"so-called instruction": Christine Frederick, *Selling Mrs. Consumer* (New York: Business Bourse, 1929), p. 186.

165 "Fancy cooking cutlets": Maud Lancaster, *Electric Cooking, Heating, Cleaning, Etc.: Being a Manual of Electricity in the Service of the Home*, ed. E. W. Lancaster (London: Constable, 1914), frontispiece.

"There is no household": A. E. Kennelly, "Electricity in the Household," in *Electricity in Daily Life: A Popular Account of the Applications of Electricity to Every Day Uses* (New York: Charles Scribner's Sons, 1891), p. 252.

"A traveler will find": "Electricity in the Household," *Scientific American*, March 19, 1904, p. 232.

"can be very delicately": Ibid.

"Even an invalid": Ibid.

166 "went after every kind": Harold Platt, interview, "Program Two: Electric Nation," in *Great Projects: The Building of America*, http://www.pbs.org/greatprojects/interviews/platt_1.html (accessed April 7, 2009).

167 "A tin can": Frederick, *Selling Mrs. Consumer*, p. 157.

168   "electricity, the unseen": Hungerford, "Night Glow of the City," p. 14.

"Woman has been": Mary Pattison, "The Abolition of Household Slavery," in *Giant Power: Large Scale Electrical Development as a Social Factor*, ed. Morris Llewellyn Cooke (Philadelphia: American Academy of Political and Social Science, 1925), p. 124.

"They have let go": H. R. Kelso, *House Furnishing Review*, July 1919, quoted in Lifshey, *The Housewares Story*, p. 289.

169   "As a matter of fact": *Ladies' Home Journal*, quoted in Barbara Ehrenreich and Deirdre English, *For Her Own Good: 150 Years of the Experts' Advice to Women* (Garden City, NY: Anchor Press, 1978), p. 135.

"We can see and feel": Frederick W. Taylor, *The Principles of Scientific Management*, 1911, Modern History SourceBook, http//www.fordham.edu/HALSALL/MOD/1911taylor.html (accessed March 26, 2006).

"The cry of the home": Pattison, "The Abolition of Household Slavery," pp. 126–27.

170   "Because we housewives": *Ladies' Home Journal*, quoted in Ehrenreich and English, *For Her Own Good*, p. 162.

171   "Rise from bed": F. Scott Fitzgerald, *The Great Gatsby* (New York: Scribner, 2004), p. 173.

"When the gas": Brian Bowers, *Lengthening the Day: A History of Lighting Technology* (Oxford: Oxford University Press, 1998), p. 132.

"In the parlor": Kennelly, "Electricity in the Household," p. 246.

"When they say": E. B. White, "Sabbath Morn," in *One Man's Meat*, enl. ed. (New York: Harper & Row, 1944), p. 51.

172   "Walk around the outside": Charles Frederick Weller, *Neglected Neighbors: Stories of Life in the Alleys, Tenements and Shanties of the National Capital* (Philadelphia: John C. Winston, 1909), pp. 10–11.

173   "The whites generally occupied": David Hajdu, *Lush Life: A Biography of Billy Strayhorn* (New York: North Point Press, 2000), p. 7.

"ironing beside": Weller, *Neglected Neighbors*, pp. 17–19.

"the perspiring woman": Ibid., pp 82–83.

174   "each day was a scuffle": Ethel Waters, with Charles Samuels, *His Eye Is on the Sparrow: An Autobiography* (Garden City, NY: Doubleday, 1951), p. 46.

"The prettiest sight": Ibid., pp. 18–19.

CHAPTER 12: ALONE IN THE DARK

175 "They are pronounced": James Agee and Walker Evans, *Let Us Now Praise Famous Men: Three Tenant Families* (Boston: Houghton Mifflin, 1988), pp. 265–66.

176 "will be a highly skilled": Quoted in Clark C. Spence, "Early Uses of Electricity in American Agriculture," *Technology and Culture* 3, no. 2 (Spring 1962): 150.
"not improbably": *Country Gentleman*, quoted ibid., p. 144.

177 "There was no quittin'": Quoted in Mary Ellen Romeo, *Darkness to Daylight: An Oral History of Rural Electrification in Pennsylvania and New Jersey* (Harrisburg: Pennsylvania Rural Electric Association, 1986), p. 13.

178 "You could milk a cow": Quoted ibid., pp. 18–19.
"Winter mornings": Quoted in Robert A. Caro, *The Years of Lyndon Johnson: The Path to Power* (New York: Alfred A. Knopf, 1982), p. 503.

179 "I would have to get": Quoted ibid., p. 505.
"You see how round": Quoted ibid.
"I have always lived": Quoted in Katherine Jellison, *Entitled to Power: Farm Women and Technology, 1913–1963* (Chapel Hill: University of North Carolina Press, 1993), p. 14.
"I got up many": Quoted in Romeo, *Darkness to Daylight*, p. 12.
"By the time": Quoted in Caro, *The Years of Lyndon Johnson*, p. 509.

180 "Our artificial light": Jimmy Carter, *An Hour Before Daylight: Memories of a Rural Boyhood* (New York: Simon & Schuster, 2001), p. 31.
"You know, you couldn't": Quoted in Romeo, *Darkness to Daylight*, p. 19.

181 "this jazz-industrial age": M. L. Wilson, quoted in Russell Lord, "The Rebirth of Rural Life, Part 2," *Survey Graphic* 30, no. 12 (December 1941), http://newdeal.feri.org/survey/sg41687.htm (accessed March 12, 2006).
"This is the test": David E. Nye, *Image Worlds: Corporate Identities at General Electric, 1890–1930* (Cambridge, MA: MIT Press, 1985), photo, insert after p. 134.

182 "The thing [the farm woman] needs": Quoted in Jellison, *Entitled to Power*, p. 13.
"We would like": Quoted ibid., p. 67.

"everything had already": Quoted in Caro, *The Years of Lyndon Johnson*, p. 512.

"the kind of oil": William T. O'Dea, *The Social History of Lighting* (London: Routledge & Kegan Paul, 1958), p. 56.

"Kerosene light": Agee and Evans, *Let Us Now Praise Famous Men*, p. 211.

183   "A blown-out electric bulb": Ibid., pp. 437–38.

"street lighting in the United States": David E. Nye, *Electrifying America: Social Meanings of a New Technology, 1880–1940* (Cambridge, MA: MIT Press, 1992), p. 140.

184   "provide a link": Quoted in Jonathan Coopersmith, *The Electrification of Russia, 1880–1926* (Ithaca, NY: Cornell University Press, 1992), p. 154.

"displayed an illuminated map": Ibid., p. 1.

"Ten years ago": Harold Evans, "The World's Experience with Rural Electrification," in *Giant Power: Large Scale Electrical Development as a Social Factor*, ed. Morris Llewellyn Cooke (Philadelphia: Academy of Political and Social Science, 1925), p. 33.

185   "the kw.h. production": Ibid., p. 36.

"far off above Manhattan": "Edison Is Buried on 52d Anniversary of Electric Light," *New York Times*, October 22, 1931, p. 1.

186   "Mr. Hoover left it": "Nation to Be Dark One Minute Tonight After Edison Burial," *New York Times*, October 21, 1931, p. 1.

CHAPTER 13: RURAL ELECTRIFICATION

187   "It is more important": *Report of the Country Life Commission: Report and Special Message from the President of the United States*, 60th Cong., 2d sess., Senate Document 705 (Spokane, WA: Chamber of Commerce, 1911), pp. 30–31, Core Historical Literature of Agriculture, http://chla.library.cornell.edu (accessed February 15, 2008).

"drive a wedge": Martha Bensley Bruère, "What Is Giant Power For?" in *Giant Power: Large Scale Electrical Development as a Social Factor*, ed. Morris Llewellyn Cooke (Philadelphia: American Academy of Political and Social Science, 1925), p. 120.

188   "When the first-of-the-month": Franklin Delano Roosevelt, quoted in Jackie Kennedy, "Seeds for America's Rural Electricity Sprouted in Diverse Power Service Territory," http://www.diversepower.com/history_heritage.php (accessed February 14, 2008).

189 "Power is really": Press conference, Franklin Delano Roosevelt, Warm Springs, GA, November 23, 1934, http://georgiainfo.galileo .usg.edu/FDRspeeches/FDRspeech34-2.htm (accessed July 9, 2009).

190 "Now the Alcorn County": Ibid.
"There must have been": David E. Lilienthal, *The Journals of David E. Lilienthal*, vol. 1, *The TVA Years, 1939–1945* (New York: Harper & Row, 1964), p. 52.

191 "full even without": Eleanor Buckles, *Valley of Power* (New York: Creative Age Press, 1945), p. 18.
"And since there wasn't": Quoted in Michael J. McDonald and John Muldowny, *TVA and the Dispossessed: The Resettlement of Population in the Norris Dam Area* (Knoxville: University of Tennessee Press, 1982), p. 40.

192 "I guess they felt": John Rice Irwin, quoted ibid., p. 57.
"And the people": Ibid.

194 "From all this": Cranston Clayton, "The TVA and the Race Problem," *Opportunity: Journal of Negro Life* 12, no. 4 (April 1934): 111, http://newdeal.feri.org/search_details.cfm?link=http://newdeal.feri .org/opp/opp34111.htm (accessed March 12, 2006).

195 "A malaria-ridden": Buckles, *Valley of Power*, p. 123.
"We were all": John Carmody, quoted in Dr. Tom Venables, "The Early Days: A Visit with John M. Carmody," *Rural Electrification* 19, no. 1 (October 1960): 20.

197 "Initially . . . the REA": Katherine Jellison, *Entitled to Power: Farm Women and Technology, 1913–1963* (Chapel Hill: University of North Carolina Press, 1993), p. 98.
"Construction crews . . . have dug": *Rural Electrification on the March* (Washington, DC: Rural Electrification Administration, July 1938), p. 7.
"An Indiana woman": Richard A. Pence, ed., *The Next Greatest Thing: 50 Years of Rural Electrification in America* (Washington, DC: National Rural Electric Cooperative Association, 1984), p. 95.

198 "In Virginia, a co-op": Ibid., p. 88.

199 "I had gotten": Quoted in Mary Ellen Romeo, *Darkness to Daylight: An Oral History of Rural Electrification in Pennsylvania and New Jersey* (Harrisburg: Pennsylvania Rural Electric Association, 1986), p. 61.
"We had a large": Jimmy Carter, *An Hour Before Daylight: Memories of a Rural Boyhood* (New York: Simon & Schuster, 2001), p. 32.

"The day we got": Quoted in *Rural Lines — USA: The Story of Cooperative Rural Electrification*, rev. ed. (N.p.: U.S. Department of Agriculture, 1981), p. 14.

200 "They report that": Quoted in Romeo, *Darkness to Daylight*, p. 68.

"I think the best day": Jimmy Carter, quoted in *Rural Lines — USA*, p. 12.

"We felt like": Quoted in Romeo, *Darkness to Daylight*, p. 100.

"Electricity changed the country": Quoted ibid.

201 "was wonderful": Quoted ibid., p. 55.

"I'll never forget": Quoted ibid.

For those in cities: Edward Hopper's painting is titled *Nighthawks* (1942).

"That light in the kitchen": Quoted in Romeo, *Darkness to Daylight*, pp. 55–56.

"Some of them wanted": Quoted ibid., p. 58.

202 "I've seen this happen": Quoted ibid., p. 56.

"Buried here May 3": Photo, ibid., p. 59.

"What is electricity": Hurst Mauldin and William A. Cochran Jr., *Electricity for the Farm* (N.p.: Alabama Power Company, 1960), p. 1.

203 "All this pushbutton stuff": Quoted in McDonald and Muldowny, *TVA and the Dispossessed*, p. 30.

"To a farm girl": Quoted in Jellison, *Entitled to Power*, p. 149.

"I would never": Quoted in *Rural Electrification on the March*, p. 70.

The advancing electric lines: John Bisbee, conversation with the author, August 2008.

CHAPTER 14: COLD LIGHT

205 "Practically every illuminant": E. Newton Harvey, "Cold Light," *Scientific Monthly*, March 1931, p. 270.

"Today we are producing": Charles Steinmetz, quoted in "Scientists Racing to Find Cold Light," *New York Times*, April 24, 1922, p. 5.

"A 60-watt bulb": Paul W. Keating, *Lamps for a Brighter America: A History of the General Electric Lamp Business* (New York: McGraw-Hill, 1954), p. 5.

206 "What a preposterous": "Nikola Tesla and His Work," *New York Times*, September 30, 1894, p. 20.

"Here you have": Harvey, "Cold Light," p. 272.

207 "At sunset the firefly": Walter Hough, *Fire as an Agent in Human Culture*, bulletin no. 139, Smithsonian Institution (Washington, DC: Government Printing Office, 1926), pp. 197–98.

208 "There were at first": Quoted ibid., p. 196.

"I think it is possible": Steinmetz, quoted in "Scientists Racing to Find Cold Light," p. 5.

211 "The road to Tomorrow": E. B. White, "The World of Tomorrow," in *Essays of E. B. White* (New York: Harper & Row, 1977), p. 111.

"Only selected parts": Hugh O'Connor, "Science at the World's Fair — Rise of the Illuminating Engineer," *New York Times*, June 11, 1939, p. D4.

"As night fell": Helen A. Harrison, "The Fair Perceived: Color and Light as Elements in Design and Planning," in *Dawn of a New Day: The New York World's Fair, 1939/40* (New York: New York University Press, 1980), p. 46.

"bore an uncanny resemblance": Ibid.

"Even the drabbest": Ibid., pp. 46–47.

212 "It's easy to see": Keating, *Lamps for a Brighter America*, photo, insert after p. 184.

CHAPTER 15: WARTIME: THE RETURN OF OLD NIGHT

215 "The earth grew spangled": Antoine de Saint-Exupéry, *Night Flight*, trans. Stuart Gilbert (New York: Century, 1932), p. 8.

216 "Experience has shown": Quoted in Williamson Murray, *War in the Air, 1914–1945* (London: Cassell, 1999), pp. 69–70.

217 Those in the steel industry: Terence H. O'Brien, *Civil Defense* (London: Her Majesty's Stationery Office and Longmans, Green, 1955), p. 229n.

219 Without streetlights: Ibid., p. 322.

220 "From different angles": Vera Brittain, *England's Hour* (New York: Macmillan, 1941), pp. 213–14.

221 October 15 saw: Angus Calder, *The People's War: Britain, 1939–45* (New York: Pantheon Books, 1969), p. 168.

"Whatever part of London": Brittain, *England's Hour*, p. 121.

"the clatter of little": Calder, *The People's War*, p. 170.

"Yet another raider": Graham Greene, *The Ministry of Fear*, in *3 by Graham Greene* (New York: Viking Press, 1948), p. 19.

"Over the night": Brittain, *England's Hour*, p. 113.

"[They] had taken over": Henry Moore and John Hedgecoe, *Henry Moore: My Ideas, Inspiration and Life as an Artist* (London: Collins & Brown, 1999), p. 170.

222 "And amid the grim": Ibid.

"a pure and curious": Elizabeth Bowen, quoted in Calder, *The People's War*, p. 173.

"What surrounded us": Hans Erich Nossack, *The End: Hamburg, 1943*, trans. Joel Agee (Chicago: University of Chicago Press, 2004), pp. 37–38.

223 "There is said": "Mission Develops U.S. Civil Defense," *New York Times*, February 14, 1941, p. 6.

224 "Get off the streets": "Fog Blanket Aids in Blackout Test of All Manhattan," *New York Times*, May 23, 1942, p. 1.

"The crowds melted into,": Ibid., pp. 1–2.

225 "As the lights came on": Ibid., p. 2.

"For every undraped window": "London Lights Up Somewhat Hesitantly; War Habits Persist After End of Blackout," *New York Times*, April 24, 1945, p. 19.

"The few householders": Ibid.

CHAPTER 16: LASCAUX DISCOVERED

226 "I made myself": Marcel Ravidat, quoted in Mario Ruspoli, *The Cave of Lascaux: The Final Photographs* (New York: Harry N. Abrams, 1987), p. 188.

227 "We raised the lamp": Ibid.

"Like a trail": Ibid., p. 189.

Scientists and archaeologists: The names of the chambers of the Lascaux Cave and the figures in them are from Norbert Aujoulat, *Lascaux: Movement, Space, and Time*, trans. Martin Street (New York: Harry N. Abrams, 2005), p. 30.

228 "in a prairie": Ibid., p. 191.

"The lights were never": Ruspoli, *The Cave of Lascaux*, pp. 180, 182, 183.

PART IV

231 "Science tells us": Vladimir Nabokov, *Pale Fire* (London: Weidenfeld & Nicolson, 1962), p. 193.

"Nothing, storm or flood": Ralph Ellison, *Invisible Man* (New York: Random House, 1995), p. 7.

CHAPTER 17: BLACKOUT, 1965

233 ". . . we have built": Robinson Jeffers, "The Purse-Seine," in *Rock and Hawk: A Selection of Shorter Poems*, ed. Robert Hass (New York: Random House, 1987), p. 191.

By 1960, on the twenty-fifth: Statistics on Rural Electrification are from *The Rural Electric Fact Book* (Washington, DC: National Rural Electric Cooperative Association, 1960), pp. 3, 56.

234 "It is scarcely": R. R. Bowker, ed., "Electricity," no. 12 in The Great American Industries series, *Harper's*, October 1896, p. 728.

"In times of normal": Paul L. Montgomery, "And Everything Was Gone," in *The Night the Lights Went Out*, ed. A. M. Rosenthal (New York: New American Library, 1965), p. 19.

235 "A slight variation": John Noble Wilford and Richard F. Shepard, "Detective Story," in *The Night the Lights Went Out*, p. 84.

"is like a game": Matthew L. Wald, Richard Pérez-Peña, and Neela Banerjee, "The Blackout: What Went Wrong; Experts Asking Why Problems Spread So Far," *New York Times*, August 16, 2003, http://www.nytimes.com (accessed May 3, 2007).

236 "Because the relay": Wilford and Shepard, "Detective Story," p. 86.

237 "In the New York State system": Donald Johnston, "The Grid," in *The Night the Lights Went Out*, p. 75.

238 "I don't know why": Quoted in Montgomery, "And Everything Was Gone," p. 23.

"'The Chinese'": A. M. Rosenthal, "The Plugged-in Society," in *The Night the Lights Went Out*, p. 11.

"through the minds": Ibid., p. 14.

239 "I could see": Quoted in Montgomery, "And Everything Was Gone," p. 20.

"like hamsters": Quoted ibid., p. 24.

"glided down more": "The Talk of the Town: Notes and Comment," *The New Yorker*, November 20, 1965, p. 45.

"As usual New Yorkers": Rosenthal, "The Plugged-in Society," p. 12.

240 "The more efficient": Wolfgang Schivelbush, quoted in David E.

Nye, *Technology Matters: Questions to Live With* (Cambridge, MA: MIT Press, 2007), p. 163.

"It was a beautiful": Quoted in Paul L. Montgomery, "The Stricken City," in *The Night the Lights Went Out*, pp. 37–38.

241 "We still knew nothing": "The Talk of the Town," November 20, 1965, p. 44.

"as if the darkness": Ibid., p. 43.

"the men, working without": Montgomery, "The Stricken City," p. 44.

"Two matches, carefully tended": "The Talk of the Town," November 20, 1965, p. 45.

242 "The moonlight lay": Ibid., p. 46.

244 "The turbine generators": William E. Farrell, "The Morning After," in *The Night the Lights Went Out*, p. 66.

"Unfortunately many": Gordon D. Friedlander, "The Northeast Power Failure — a Blanket of Darkness," *IEEE Spectrum*, February 1966, p. 66.

245 "As power became available": *Report to the President by the Federal Power Commission on the Power Failure in the Northeastern United States and the Province of Ontario on November 9–10, 1965*, December 6, 1965, p. 29, http://www.blackout.gmu.edu/archive/pdf/fpc_65.pdf.

"New York Cancelled": Bernard Weinraub, "From Abroad: Smiles, Sneers, and Disbelief," in *The Night the Lights Went Out*, p. 119.

"Ralph Morse, who had": George P. Hunt, "Trapped in a Skyscraper," *Life*, November 19, 1965, p. 3.

246 "Everybody recognizes everybody": Farrell, "The Morning After," p. 65.

The subsequent Federal Power: *Report to the President*, pp. 43–45.

248 "We are in much worse": "The Talk of the Town: Notes and Comment," *The New Yorker*, August 15, 1977, p. 15.

"The end came": Russell Baker, quoted in Bernard Weinraub, "Bewitched and Bewildered," in *The Night the Lights Went Out*, pp. 124–25.

CHAPTER 18: IMAGINING THE NEXT GRID

250 "Regard the light": Dan Flavin, "'. . . in Daylight or Cool White': An Autobiographical Sketch," *Artforum*, December 1965, p. 24.

251 "Permanence just defies": "Dan Flavin Interviewed by Tiffany Bell, July 13, 1982," in *Dan Flavin: The Complete Lights, 1961–1996*, ed. Michael Govan and Tiffany Bell (New Haven, CT: Dia Art Foundation / Yale University Press, 2004), p. 199.

"Oil had become": Daniel Yergin, *The Prize: The Epic Quest for Oil, Money, and Power* (New York: Simon & Schuster, 1991), p. 588.

252 "It's very sad": Quoted in "The Talk of the Town: Other Lights," *The New Yorker*, December 10, 1973, p. 40.

253 "This winter as the nation": Jonathan Schell, "The Talk of the Town: Notes and Comment," *The New Yorker*, December 10, 1973, p. 37.

"Night's coming was": Baron Wormser, *The Road Washes Out in Spring: A Poet's Memoir of Living Off the Grid* (Hanover, NH: University Press of New England, 2006), p. 9.

"A few guests": Ibid., p. 11.

254 "Light did not materialize": Ibid., p. 10.

"We simply must balance": Jimmy Carter, speech, April 18, 1977, "Primary Sources: The President's Proposed Energy Policy," *American Experience*, http://www.pbs.org/wgbh/amex/carter/filmmore/ps_energy.html (accessed May 2, 2008).

256 "We're working to create": Jeffrey Skilling, quoted in Steven Johnson, "New New Power Business: Inside 'Energy Alley,'" *Frontline*, http://www.pbs.org/wgbh/pages/frontline/shows/blackout/traders/inside.html (accessed December 2, 2008).

258 "You probably couldn't": Jeffrey Skilling, quoted in Bethany McLean and Peter Elkind, *The Smartest Guys in the Room: The Amazing Rise and Scandalous Fall of Enron* (New York: Penguin Books, 2004), p. 281.

"You know what": Ibid.

"They should just": Quoted in "Enron Trader Conversations: 'Powerex and Bonneville . . . ,'" Ex. SNO — 224, pp. 5–6, *Seattle Times*, February 4, 2005, http://seattletimes.nwsource.com/html/localnews/2001945474_webenronaudio02.html (accessed September 27, 2009).

259 "I used contemporary": Steven Watt, conversation with the author, October 2008.

"the most significant": U.S. Department of Energy, *The Smart Grid: An Introduction* (Washington, DC: U.S. Department of Energy, n.d.), p. 5.

261 "Imagine all the south-facing": Bill McKibben, *Deep Economy: The Wealth of Communities and the Durable Future* (New York: Times Books, 2007), p. 145.

262 "Energy is at the core": Richard E. Smalley, testimony to the Senate Committee on Energy and Natural Resources, Hearing on Sustainable, Low Emission, Electricity Generation, April 27, 2004, http://www.energybulletin.net/note/249 (accessed October 18, 2008).

263 "From about 1990": Brian Bowers, *Lengthening the Day: A History of Lighting Technology* (Oxford: Oxford University Press, 1998), p. 190.

264 "It took me almost": Gavin Hudson, "Korea Shines for Compact Fluorescent Use," *EcoWorldly*, January 9, 2008, http://ecoworldly .com/2008/01/09/brilliant-asia-cfls-are-turning-korea-on (accessed March 11, 2009).
"You wake up": Quoted in "Making the Switch (or Not)," *New York Times*, January 10, 2008, p. D6.
"No, the light quality": Ibid.

265 "Do not use": "What If I Accidentally Break a Fluorescent Lamp in My House?" Maine Department of Environmental Protection, Bureau of Remediation and Waste Management, http://www.maine .gov/dep/rwm/homeowner/cflbreakcleanup.htm (accessed April 11, 2009).

266 "The candle does not": Gaston Bachelard, *The Flame of a Candle*, trans. Joni Caldwell (Dallas: Dallas Institute Publications, 1988), p. 37.

267 "The candle will burn out": Ibid.
"the unique combination": "Reproduction Light Bulbs," Rejuvenation: Classic American Lighting & House Parts, http://www.rejuv enation.com/templates/collection.phtml?accessories=Reproduction %20Bulbs (accessed May 3, 2009).

CHAPTER 19: AT THE MERCY OF LIGHT

270 "I wanted to investigate": Michel Siffre, *Beyond Time: The Heroic Adventure of a Scientist's 63 Days Spent in Darkness and Solitude in a Cave 375 Feet Underground*, ed. and trans. Herma Briffault (London: Chatto & Windus, 1965), p. 25.
"This morning I was": Ibid., pp. 154–55.

271 "I emerged": Michel Siffre, "Six Months Alone in a Cave," *National Geographic*, March 1975, p. 428.

"Forty-second awakening": Siffre, *Beyond Time*, pp. 166, 181–82.

"I underestimated": Ibid., pp. 222, 225.

272 "meaning that most": Warren E. Leary, "Feeling Tired and Run Down? It Could Be the Lights," *New York Times*, February 8, 1996, http://www.nytimes.com (accessed August 9, 2007).

"Every time we turn on": Dr. Charles Czeisler, quoted ibid.

273 Divided sleep: See A. Roger Ekirch, "Sleep We Have Lost: Preindustrial Slumber in the British Isles," *American Historical Review* 106, no. 2 (April 2001), http://www.historycooperative.org/journals /ahr/106.2/ahooo343.html (accessed July 4, 2007).

274 "There is one stirring": Robert Louis Stevenson, "A Night Among the Pines," in *"Travels with a Donkey in the Cévennes" and "The Amateur Emigrant"* (London: Penguin Books, 2004), pp. 56–57.

"slept only about an hour": Natalie Angier, "Modern Life Suppresses an Ancient Body Rhythm," *New York Times*, March 14, 1995, http://www.nytimes.com (accessed August 9, 2007).

275 "We think Thomas Edison": Czeisler, quoted in Leary, "Feeling Tired and Run Down?"

"Everything which decreases": "Edison's Prophesy: A Duplex, Sleepless, Dinnerless World," *Literary Digest*, November 14, 1914, p. 966.

276 "Offshore hydrocarbon platforms": William A. Montevecchi, "Influences of Artificial Light on Marine Birds," in *Ecological Consequences of Artificial Night Lighting*, ed. Catherine Rich and Travis Longcore (Washington, DC: Island Press, 2006), p. 100.

"Many nocturnal species": Paul Beier, "Effects of Artificial Night Lighting on Terrestrial Mammals," in *Ecological Consequences of Artificial Night Lighting*, pp. 32–33.

277 "exploring new habitat": Ibid., p. 34.

278 "on misty and foggy": Sidney A. Gauthreaux Jr. and Carroll G. Belser, "Effects of Artificial Night Lighting on Migrating Birds," in *Ecological Consequences of Artificial Night Lighting*, p. 77.

"The habit of feeding": Jens Rydell, "Bats and Their Insect Prey at Streetlights," in *Ecological Consequences of Artificial Night Lighting*, p. 43.

279 "humans are changing": Bryant W. Buchanan, "Observed and Poten-

tial Effects of Artificial Lighting on Anuran Amphibians," in *Ecological Consequences of Artificial Night Lighting*, p. 215.

"as they have fallen": David Ehrenfeld, "Night, Tortuguero," in *Ecological Consequences of Artificial Night Lighting*, p. 138.

281 "A light break": Winslow R. Briggs, "Physiology of Plant Responses to Artificial Lighting," in *Ecological Consequences of Artificial Night Lighting*, p. 401.

"the thousands of little": Michael Pollak, "'Towers of Light' Awe," *New York Times*, October 10, 2004, http://www.nytimes.com (accessed October 13, 2008).

"Some people thought": Ibid.

CHAPTER 20: MORE IS LESS

282 "At the second match": Robert Louis Stevenson, "Upper Gévaudan," in *"Travels with a Donkey in the Cévennes" and "The Amateur Emigrant"* (London: Penguin Books, 2004), p. 30.

"One night I went": Vincent van Gogh to Theo van Gogh, letter 499, in *The Complete Letters of Vincent van Gogh*, vol. 2 (Greenwich, CT: New York Graphic Society, 1959), p. 589.

283 "has overpopulated": Charles Whitney, "The Skies of Vincent van Gogh," *Art History* 9, no. 3 (September 1986): 353.

"*I should be desperate*": Vincent van Gogh to Theo van Gogh, letter 418, in *The Complete Letters of Vincent Van Gogh*, vol. 2, p. 401.

284 "brilliant with its own": Ovid, *Metamorphoses*, quoted in Bart J. Bok and Priscilla F. Bok, *The Milky Way* (Cambridge, MA: Harvard University Press, 1981), p. 1.

"emergency organizations": Terence Dickinson, *NightWatch: A Practical Guide to Viewing the Universe* (Buffalo, NY: Firefly Books, 1998), p. 47.

285 "About one-tenth": P. Cinzano, F. Falchi, and C. D. Elvidge, *The First World Atlas of the Artificial Night Sky Brightness*, abstract, p. 1, http://www.inquinamentoluminoso.it/cinzano/download/0108052.pdf (accessed June 8, 2009).

"Surely it is": Galileo Galilei, *The Starry Messenger*, p. 1, http://www.bard.edu/admission/forms/pdfs/galileo.pdf (accessed June 8, 2009).

"Here we have": Ibid., p. 14.

"is not robed": Ibid., p. 1.

286 "With the aid": Ibid., p. 10.
"Many astronomers thought": Ronald Florence, *The Perfect Machine: Building the Palomar Telescope* (New York: HarperCollins, 1994), p. 106.

288 "The 200-inch": Edwin Hubble, quoted ibid., p. 395.
"Astronomy is an incremental": Florence, *The Perfect Machine*, p. 404.
"It's like I'm looking": Quoted in Mari N. Jensen, "Light Pollution in Tucson," *Tucson Citizen*, August 21, 2001, http://www-kpno.kpno .noao.edu/pics/lighting/tucsoncitizen_8_21_01light.html (accessed October 14, 2008).

289 "When you take": Dave Kornreich, "How Does Light Pollution Affect Astronomers?" *Curious About Astronomy? — Ask an Astronomer,* April 1999, p. 1, http://curious.astro.cornell.edu/question.php?num ber=194 (accessed September 18, 2007).

290 "Light traversing a path": Bob Mizon, *Light Pollution: Responses and Remedies* (London: Springer-Verlag, 2002), p. 34.
"the city lights": Kornreich, "How Does Light Pollution Affect Astronomers?" p. 2.

291 "is equivalent to": Richard Preston, *First Light: The Search for the Edge of the Universe* (New York: Atlantic Monthly Press, 1987), p. 24.

292 "Then Humankind was born": Ovid, *Metamorphoses*, trans. A. S. Kline, 1.68–88, http://etext.virginia.edu/latin/ovid/trans/Metamorph .htm (accessed June 29, 2009).

CHAPTER 21: THE ONCE AND FUTURE LIGHT

293 "The spiritual instant": Henri Focillon, *The Life of Forms in Art*, trans. Charles Beecher Hogan and George Kubler (New York: Zone Books, 1992), p. 152.

294 "The *Home Office*": Bob Mizon, *Light Pollution: Responses and Remedies* (London: Springer-Verlag, 2002), p. 61.
"a car storage area": Ibid.
How complex the relation: For information on the study of Chicago's alleyways, see *The Chicago Alley Lighting Project: Final Evaluation Report*, April 2000, http://www.icjia.state.il.us/public/pdf/ResearchRe ports (accessed June 8, 2009).

295 "Yes, my tent became": Michel Siffre, *Beyond Time: The Heroic Adventure of a Scientist's 63 Days Spent in Darkness and Solitude in a Cave 375 Feet Underground*, ed. and trans. Herma Briffault (London: Chatto & Windus, 1965), pp. 99–100.

296 "A growing number": "Sustainability, Urban Planning, and What They Mean to Dark Skies," *Newsletter of the International Dark-Sky Association*, http://www.darksky.org/news/newsletters/60-69/nl66_fea .html (accessed May 23, 2007).

297 "a new city of light": Hollister Noble, "New York's Crown of Light," *New York Times*, February 8, 1925, p. SM2.

298 "The tall tower": Ken Belson, "Efficiency's Mark: City Glitters a Little Less," *New York Times*, November 2, 2008, http://www.ny times.com (accessed March 11, 2009).

299 "Unfortunately most of today's": John E. Bortle, "Introducing the Bortle Dark-Sky Scale," *Sky & Telescope*, February 2001, p. 126.

300 "There's a good part": Quoted in Dave Caldwell, "Dark Sky, Bright Lights," *New York Times*, September 14, 2007, p. F10.

301 "When my mother": Alhassan Sillah, "Fuel for Thought in Guinea," *BBC News*, http://newsvote.bbc.co.uk/mpapps/pagetools/print/news .bbc.co.uk/2/h (accessed March 14, 2009).
"I hardly ever": Ibid.
"I used to study": Rukmini Callimachi, "Kids in Guinea Study Under Airport Lamps," *Washington Post*, http://www.washingtonpost.com /wp-dyn/content/article/2007/07/19 (accessed March 14, 2009).

302 "Working in the so-called": Sheila Kennedy, quoted in "Light unto the Developing World," *Miller-McCune Magazine*, http://www .miller-mccune.com/article/light-unto-the-developing-world (accessed December 13, 2008).
"Instead of a centralized": Kennedy, quoted in "Energizing the Household Curtain," JumpIntoTomorrow.com, http://www.jumpin totomorrow.com/template/index/php?tech=82 (accessed December 14, 2008).

EPILOGUE

305 For further information on ways to reduce light pollution, see International Dark-Sky Association, http://www.darksky.org, and Fatal Light Awareness Program (FLAP), http://www.flap.org.

INDEX